Climate Change in Canada

Rodney White

ISSUES IN CANADA

OXFORD
UNIVERSITY PRESS

OXFORD
UNIVERSITY PRESS

8 Sampson Mews, Suite 204, Don Mills, Ontario M3C 0H5
www.oupcanada.com

Oxford University Press is a department of the University of Oxford.
It furthers the University's objective of excellence in research, scholarship,
and education by publishing worldwide in

Oxford New York
Auckland Cape Town Dar es Salaam Hong Kong Karachi Kuala Lumpur Madrid
Melbourne Mexico City Nairobi New Delhi Shanghai Taipei Toronto

With offices in
Argentina Austria Brazil Chile Czech Republic France Greece
Guatemala Hungary Italy Japan Poland Portugal Singapore
South Korea Switzerland Thailand Turkey Ukraine Vietnam

Oxford is a trade mark of Oxford University Press in the UK and in certain other countries

Published in Canada by Oxford University Press

First Published 2010

Library and Archives Canada Cataloguing in Publication

White, Rodney
Climate change in Canada / Rodney White.

(Issues in Canada)
Includes bibliographical references and index.

ISBN 978-0-19-543060-8

1. Climatic changes—Canada. 2. Climatic changes—Government
policy—Canada. I. Title. II. Series: Issues in Canada.

QC981.8.C5W52 2009 363.738'740971 C2008-907287-1

Cover image: Courtesy of Andy Clark, PhotoSensitive

Printed and bound in Canada.

1 2 3 4 — 13 12 11 10

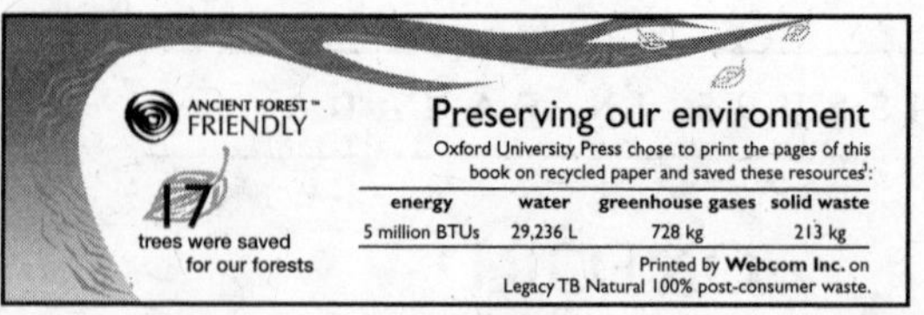

[1]Estimates were made using the Environmental Defense Paper Calculator.

Contents

List of Tables

List of Figures

Acknowledgements

In a book of this breadth the author inevitably draws on the assistance and insights of many friends and colleagues over a long period of time. For *Climate Change in Canada* this is certainly the case. To identify all the individuals from whom I have learned something that pertains to this book would require another book in itself. So my acknowledgements will be more institutional than individual.

I first came to the University of Toronto to teach in the Department of Geography in 1973 and I have made most of my career here, in which time I have enjoyed the company and stimulation of many colleagues. In 1980 I was cross-appointed to the Institute for Environmental Studies, where I was ideally placed to learn from colleagues in many disciplines other than my own. In the 1980s the Institute became part of the international research network for Analysing Biospheric Change and it was probably under the aegis of the ABC Project that I became attuned to the kind of global overview that helps us to come to terms with the breadth of the climate change challenge.

In the 1990s the Institute formed a partnership with the Adaptation and Impacts Research Group (AIRG) of Environment Canada and it was with AIRG that we set up a Round Table on Natural Hazards with members of the insurance industry. It was at that time that Environment Canada was trying to replace the "global warming" message of climate change with the idea that we should expect more "extreme events" as part of the changing climate. The insurers knew all about the financial implications of extreme events.

It was at this time that the field of environmental finance was beginning to emerge, mostly due to the work of the US Environmental Protection Agency in the use of markets to reduce acid rain. As explained in this book (and elsewhere), it was the success of the acid rain reduction program that encouraged the UNFCCC to employ market mechanisms to begin to control GHG emissions and hence take an important step toward the creation of the carbon market.

Although Canada has lagged behind European initiatives on GHG reduction, we have a very active business community in Canada that has been building on international experience in the development of carbon finance. At the University of Toronto we formed an Environmental Finance Advisory Committee with the support of members of this community, to offer workshops on the role of markets in environmental management. I have certainly learned a great deal from these colleagues in the business world that I would not have found at the university.

The opportunity to write this book came my way thanks to Joe Desloges (then Chair of the Department of Geography) and Lorne Tepperman, who is editing this series for Oxford University Press Canada. For this opportunity I am very grateful. I also wish to thank Jennie Rubio, my editor at OUP, for her encouragement and patience—very valuable attributes for bringing a book to life. I should also thank Sonia Labatt—my co-author on two previous books—for her interest and encouragement in this enterprise and for providing me with a copy of Nicholas Stern's book, *The Global Deal*, at a critical moment.

Most of all I thank my wife Sue for sustaining me through another book at a very busy time in our lives. I hope it will be useful in some modest measure in helping our successors sort out the mess. For Alyson, Kathryn, and Tim, and their generation, and for Isabelle and her generation, I hope the story will have a happy ending.

Rodney White, Toronto, June 2009.

Acronyms and Formulae

The world of climate change is infested by acronyms, a condition I call "acronymitis." In this book I have tried to keep them to the absolute minimum.

AIG	American Insurance Group
BP	British Petroleum
CCME	Canadian Council of Ministers of the Environment
CCS	Carbon capture and storage
CCX	Chicago Climate Exchange
CDM	Clean Development Mechanism, a mechanism under the Kyoto Protocol
CDP	Carbon Disclosure Project
CEO	Chief Executive Officer
CER	Certified Emission Reduction (produced by the CDM)
CH_4	Methane, or natural gas—a greenhouse gas
CHP	Combined Heat and Power
CIA	US Central Intelligence Agency
CO_2	Carbon dioxide—a greenhouse gas
COP	Conference of the Parties to the UNFCCC
ENGO	Environmental NGO

EPA	US Environmental Protection Agency
ETS	EU Emissions Trading Scheme
EU	European Union
GHG	Greenhouse gas
GRI	Global Reporting Initiative
HEP	Hydroelectric power
HFCs	Hydrofluorocarbons—a class of greenhouse gases
IBC	Insurance Bureau of Canada
ICLEI	International Council for Local Environmental Initiatives
IPCC	Intergovernmental Panel on Climate Change
ISO	International Organization for Standardization
NAIC	US National Association of Insurance Commissioners
NGO	Non-governmental organization
NIMBY	"Not In My Backyard" syndrome
N_2O	Nitrous oxide—a greenhouse gas
NO	Nitric oxide
NO_2	Nitrogen dioxide
NO_x	A collective term for NO and NO_2
NRTEE	National Round Table on the Environment and the Economy
PFCs	Perfluorocarbons—a class of greenhouse gases
RGGI	US Regional Greenhouse Gas Initiative
SF_6	Sulphur hexafluoride—a greenhouse gas
SO_2	Sulphur dioxide
UNFCCC	United Nations Framework Convention on Climate Change
WCI	US Western Climate Initiative

Chapter 1

Climate Change

What is happening in Canada?

In Canada, it seems that we've only recently begun talking about what we can do about global warming. These discussions occur mostly at individual, municipal, and provincial levels, with federal leadership still sorely lacking in the present Conservative (and previous Liberal) government.

Andrew Weaver, *Keeping Our Cool: Canada in a Warming World* (2008)

Canadians ... do not fully appreciate how spectacularly our governments have dithered and operated at cross-purposes on this matter for a length of time that is now coming up on two decades.

Robert Paehlke, *Some Like it Cold* (2008)

Overview

Over the last twenty years we have seen the emergence of a growing scientific consensus that global warming is underway. It is now widely accepted that climate change will have significant—and mostly negative—consequences for humanity.

Far from being immune to these consequences, Canada is particularly vulnerable.

The warming is being driven by human activities; the main ones are the emission of greenhouse gases (GHGs) and land-use change, especially deforestation. The principal GHG is carbon dioxide (CO_2), which is emitted as a by-product of the combustion of fossil fuels such as coal, oil, and natural gas. Along with water vapour, GHGs trap the heat that is radiated from the surface of the earth, preventing it from reflecting back into the atmosphere. The concentration of CO_2 in the atmosphere now is roughly 50 percent higher than it was before the Industrial Revolution.

A key factor here is the time that GHG molecules remain in the atmosphere, known as their "residence time." The residence time of CO_2 is more than 100 years. This means that we have created a long-term problem not just for ourselves, but also for our grandchildren and our great-grandchildren. The other important consequence of this long residence time is that the gas has plenty of time to become "well-mixed" in the atmosphere, before it is eventually re-absorbed into the oceans or on land. This means that greenhouse gas emissions from every country in the world get mixed together. So climate change—unlike poverty or water shortages—cannot be consigned to "over there," and hence "not our problem." It is everybody's problem. Without worldwide agreement it cannot be solved.

The warming of the lower atmosphere will increase the rate of evaporation of water vapour from the ground, from forests, and from water bodies. It will produce more intense rainfall events and also more serious droughts. Impacts will vary greatly from region to region. Within Canada the most serious water shortages will be in the Prairies—which are already prone to drought. The warming will also lead to more extreme weather events, such as downpours, hail, and tornadoes, as well as smog and heat alerts in cities. It may lead to more intense, and more frequent, hurricanes. The warming of the lower atmosphere will warm the surface waters of the ocean, which in turn will expand and produce a rise in sea level. This will increase as glaciers and icefields melt.

The warming and the changes associated with it have been widely known to scientists since the 1980s. Eventually, their concerns led to the 1992 Earth Summit in Rio de Janeiro. The summit produced the United Nations Framework Convention on Climate Change (UNFCCC), which in turn spawned the Kyoto Protocol (1997). The Protocol was ratified by the Canadian government in 2002, with a commitment to reduce Canada's GHG emissions to 6 percent below the 1990 baseline by end of the First Compliance Period (2008–2012).

But in spite of actions to reduce GHGs by Canadian municipalities, businesses, and voluntary groups, political leadership from provincial governments and the federal government has been extremely weak. Most senior governments (i.e. federal and provincial) have produced plans and made numerous promises, but as of May 2009 only the governments of British Columbia, Alberta, and Quebec have enacted measure. Not surprisingly, Canada's GHG levels have continued to rise; we are now 30 percent *over* our 1990 baseline.

This short book will consider why Canadians cannot afford to ignore climate change. In addition to the general changes in the climate discussed above, Canada faces some additional issues that require urgent resolution. First, the most rapid warming of the atmosphere will occur in higher latitudes—that is, near the poles. This has major implications for the traditional economy and culture of Canada's Aboriginal people. These effects are apparent in changing animal habitats (including fish) and transportation. The Aboriginal population is already marginalized in the Canadian economy; climate change will make things worse.

A further effect of the rapid warming in the Arctic is the reduction in sea ice. This will make resources, including oil and gas, more "accessible"—in other words, more "exploitable." New leases are already being sold. The opening of access lanes to shipping will also raise issues of sovereignty in the Arctic, which were not resolved even before climate change began to alter the map. We would be very naive to assume that Canada's claim to sovereignty in the Arctic will not be contested by the United States, Russia, and other circumpolar nations.

A third major issue is the development of the oil sands, principally in Alberta and Saskatchewan. Compared with conventional oil deposits, the extraction of these energy resources requires a great deal of energy and water. This is one reason—along with the recent development of offshore oilfields in Newfoundland and Labrador—why Canada's GHG emissions have increased so much since 1990. The oil sands is one very large "elephant in the room." Canadian newspapers carry stories about global warming almost every day. Paradoxically, they also carry story after story about the development of new energy resources—oil sands, Arctic, Hibernia—usually with no mention about their connection to climate change.

Then there is the return of nuclear power: for many environmentally concerned people, "nuclear" is the uninvited guest at the low-carbon party.

So why is climate change not front-and-centre in Canadian political life? By any measure, we are choosing our future (and our descendants' future) right now.

Myths about Climate Change

Every country and culture clings to some comforting myths. In the climate change context, many Canadians believe the following:

- Canada is, for the most part, a very cold country; a bit of warming might be a good thing.
- Canada has a lot of water. As you can see from any map, we have many huge lakes, so our water supply is not vulnerable to climate change.
- Canada is a very big country, yet it is home to only some thirty-three million people. Therefore, we have lots of resources and lots of environmental resilience.
- Canadians are well known for taking environmental issues seriously.
- Canada's current greenhouse gas emissions are only 2 percent of the world's total; so any reductions Canada might make are irrelevant.
- Climate change is a slowly developing issue. We still have plenty of time to address it.

These and other comforting assumptions will be examined throughout this book. However, it might be useful to take a brief, critical look at these six myths before we start on the longer journey.

It is true that most of Canada is very cold in the winter months and that much of the north has permanently frozen ground (or "permafrost"). However ecosystems—and human dependence on them—exist in a delicate balance with temperature, humidity, and other variables. The small temperature increase we have already experienced has been enough to melt some of the permafrost and some of the sea ice. Even if we stopped emitting greenhouse gases completely tomorrow, this warming would continue for many decades. In southern parts of the country the warming will increase the number of hot days (over 30° or even 35°C), which will result in more deaths from smog and heat stress.

A key feature of Canada on any map is the number of large lakes, creating an "illusion of plenty." But these are in a sense "fossil lakes," locations where water collected when the ice sheet melted at the end of the last ice age. These lakes are the product of an earlier hydrologic regime. The turnover of the water in the Great Lakes, for example, is about 1 percent per year. If we take off more than this, through our usage and through increased evaporation due to global warming, then lake levels will fall.

Another important factor to keep in mind is that almost all of Canada's river basins include some snow and ice, some seasonal, some "permanent." At least, this was the case until recently. The spring snowmelt is a very important part of Canada's hydrology. In some regions glaciers feed river basins throughout the spring, summer, and fall. This is especially critical in the Prairies. As the climate warms, the quantity of water originating from melting snow will diminish, even disappear, in some southern basins. Glaciers are in retreat throughout the country.

Is Canada really so big as to be indestructible in the hands of only thirty-three million people? No. The human impact is considerable, often negative and irreversible. Anywhere you travel in the more densely populated regions, you will find evidence of endangered species, clear-cut forests, depleted groundwater, polluted rivers, and polluted airsheds.

Is there a strong Canadian commitment to the environment? There is certainly a commitment from some individuals and organizations. However, the commitment of the country *as a whole* has declined since Canadians took a leadership role at the Stockholm Conference on the Environment and the Economy (1972) and gave strong support to the Brundtland Commission in the 1980s. Despite a plethora of environmental initiatives at every level, there is a glaring lack of commitment to implement the Kyoto Protocol—currently the most important tool available to combat climate change. Canada's greenhouse gas emissions have continued to increase, both in absolute terms and per capita. Among the G8 group of the world's biggest economies, emissions from Canada and Italy have grown the most in percentage terms since 1990. If we fail to reverse this trend, other environmental efforts to preserve biodiversity and reduce pollution will most likely be nullified. As one of the richest countries in the world, Canada can hardly leave it to others to make up our deficit on this front.

It is true that currently we emit only 2 percent of the world's total GHG emissions, although our share, historically, has been much greater. But wealthier countries like Canada should demonstrate *leadership*, developing initiatives like low-carbon technology, for example.

Finally, although the changes in the climate expected over the next decade may look small, the rate at which the climate is changing is increasing compared with natural global processes. For example, the rate of temperature change we have experienced over the past 20 years is about *ten times* faster than the warming which took place at the end of the last ice age. Given that greenhouse gases stay in the atmosphere for decades, even if we stopped emitting immediately the heat trapping we have already initiated would continue. Ice caps and glaciers would continue to melt and sea levels would continue to rise. The longer we ignore the problem, the more serious the consequences will be. We need to react swiftly to contain a dangerous process that is already growing in momentum. The sooner we start, the more we can manage the damage from climate change and lower the upfront costs.

So, exactly what is the global warming process that we must try to slow down and then reverse?

Why is the Planet Warming?

The natural heat-trapping effect of water vapour, CO_2, and the other GHGs makes Earth habitable for humans and the rest of Earth's fauna and flora. This is a favourable effect of the natural greenhouse effect: without this, the average temperature on Earth would not be a comfortable +15°C, but rather a freezing –18°C. Our problem lies in the fact that human activities, especially in the industrial era, have enhanced this natural effect. We are now emitting CO_2 much faster than the rate at which it naturally returns from the atmosphere to be absorbed by the oceans and by uptake in forests and other plant life. It accumulates, adding to the atmosphere's heat-trapping capacity. Deforestation and the ploughing up of land that was formerly forest or grassland releases more CO_2. It is estimated that approximately 80 percent of the enhanced warming effect is attributable to the burning of fossil fuels and 20 percent is due to our land-use practices.

The rate at which this enhanced warming takes place is driven by human numbers and human activities. The human population continues to grow, and, while many people remain trapped in poverty, others are following the Western material path, driving cars, accumulating goods, eating more meat, and burning fossil fuels. Whereas a small farmer in Africa might emit less than one fifth of a tonne of GHGs each year, the average Canadian emits more than ten tonnes.

Positive and Negative Feedbacks

Because scientists think of the earth's climate as a system, they apply systems terminology to the kind of changes we see taking place. Some of the language they use may seem counter-intuitive, particularly their use of the adjectives "positive" and "negative." In everyday language we see "positive" as "good" and "negative" as "bad." However, in systems terms, a "positive" feedback means an enhancement of a given process, while a "negative" feedback means reversal, or slowing down. Thus the continued growth of the human population and their associated emissions of CO_2 are, in systems terms, a positive feedback for the global warming trend.

Meanwhile, within this major phenomenon many smaller sub-processes contribute to the positive feedback. For example, as the world warms, people with air conditioners are likely to use them more often, thereby contributing further to fossil fuel demand and global warming. In northern Canada we can already see the impacts of global warming on the permafrost. As the permafrost breaks up it releases methane—a potent greenhouses gas—into the atmosphere, providing another positive feedback, which further enhances the greenhouse effect. Warmer weather also encourages the spread of the pine beetle through Western forests; the beetle eventually kills the pine, which then becomes susceptible to enhanced risk of forest fires, leading to further release of CO_2 and more heat-trapping capacity in the atmosphere.

To counteract all these unwanted positive feedbacks we need to develop some powerful *negative* feedback mechanisms such as price signals that discourage people from burning fossil fuels. A very strong negative feedback could be introduced in the form of carbon "ration cards": here, each individual is issued a fixed carbon budget for the year. Such a plan has been proposed in the UK and it could someday become a reality.

The Global Context

If global warming was a purely technical problem, then humanity would probably have devised an adequate response by now. Instead it is deeply embedded in the major forces which condition the human occupancy of the planet. Consider the modern rise of the human population from about one billion in 1825 to more than six billion today. This can be attributed to the process of industrialization based on access to cheap fossil fuels. It was at that point in time when our forebears made their reluctant transition from wood to coal as their principal fuel. We altered the carbon cycle, piling up excess CO_2 in the atmosphere. Interestingly, the implications of this switch were identified at the time by French scientist, Joseph Fourier (1768–1830), although no one was listening to what was probably the most prescient warning that a scientist ever provided. It was not until the 1950s that the wisdom of the transition to fossil fuels began to come into question.

In the meantime, not only was the global population growing rapidly, many people became very rich compared with their ancestors. This increase in personal wealth directly correlates with the use of increasing amounts of energy from fossil fuels and their concomitant GHG emissions. As people in the industrial world became wealthier their diets changed to include more meat, which is much more energy-intensive than grains and vegetables. We see a similar transition taking place in China and other low-income countries today. The transition to more meat in the modern Western diet requires more energy and more water than a grain-based diet. It is no coincidence that, at this time, we witness increasing concern over GHGs, water scarcity, and rising food prices on a global scale.

Many other lifestyle changes also contribute to increasing GHG emissions. The richer diet for the wealthier countries of the world is now assembled from all over the globe; "food miles" have crept into our daily consciousness. In North America, at least, houses are getting bigger, even as family size declines. Single-person and two-person households are a growing percentage of the demographic mix throughout the industrial world. On the present trajectory, energy use will continue to rise. For example, air conditioning has not yet fully penetrated the market, even in Canada, let alone in Western Europe. Holiday travel continues to grow every year, pushing up GHG emissions from airplanes and cars.

The picture is not uniform throughout Western society. Generally, Europeans use less energy and water than North Americans. Canada has the unenviable distinction of sharing with the United States the title of "most energy-intensive society in the world." Of course, there are some good reasons for this.

The Canadian Context

For Canada as a whole, three reasons are usually offered to explain the high level of energy consumption. First, much of the country for much of the year is very cold; energy is needed for space heating. Second, the country is very large, relative to its population; people travel long distances to interact, compared with Europeans, for example. Third, much of the economy is based on resource extraction (oil and gas, coal, pulp and paper, mining, agriculture) which is more energy-intensive than an economy more dependent on manufacturing and services. All of these factors explain the high level of energy use in Canada.

However, other "household" factors contribute to an energy bottom line that exceeds the comparable figures from Europe and Japan. First is the prevalence of the detached single-family dwelling, comprising approximately 60 percent of all dwellings in Canada. This has many implications, including energy used in construction, maintenance, heating, and cooling. Furthermore, most of these homes are situated in residential-only areas, where retail outlets are sparse and population density too low to support public transit. Most of the population is dependent on cars for most trips. Second, car ownership is widespread, with 77 percent of all households owning one or more vehicles. Predictably, the modal split (between different modes of transport) for commuter travel heavily favours cars.

The GHG picture across Canada, however, is very far from uniform because the "energy mix" varies greatly from province to province. For example, 80 percent of Alberta's energy comes from gas and coal, whereas 93 percent of Manitoba's energy is from hydroelectric power. Given inter-provincial variations in population size, structure of the economy, and energy mix, GHG emissions in absolute terms vary greatly. Ontario and Alberta each produce more than 200 megatonnes of "CO_2 equivalent," Quebec, Saskatchewan, and British Columbia between 70 and 100 megatonnes, and the other provinces and territories less than 20 megatonnes each. In almost every province GHG emissions have increased since 1990.

Thus although Canadians as a whole are high energy users in the global scheme, the implications of putting a "carbon overlay" across the Canadian economy (by introducing a price for carbon, through trading carbon credits, for example) would vary significantly from province to province. Roughly speaking, the biggest "carbon challenge" will be faced by Alberta and Saskatchewan, the smallest by Quebec and Manitoba, with the others lying somewhere in between.

Whose Carbon Is It, Anyway?

Only recently has Canada (at the federal and provincial level) started coming to grips with the reality that global warming is not going away, and that we will soon have to do something to honour our Kyoto pledge, beyond sending voluminous reports to the United Nations. Three provinces—Quebec, Alberta, and BC—have already led the way by introducing GHG legislation; other provinces, as well as the federal government, have promised to follow. Many initiatives have already been taken by the private sector, municipalities, and non-governmental organizations. The phrase "carbon-constrained world" is beginning to appear in Canadian newspapers. There is no escaping the fact that the global warming situation is complicated. It is unlikely that someone is going to discover a completely painless "magic bullet" solution.

However, even as Canada begins to come to terms with the magnitude of the challenge, some difficult questions have begun to emerge. Who is responsible for what? Whose carbon is it, anyway?

This book will address these complicated questions. In the meantime we might think about the forest fires resulting from the pine beetle infestation. Is the CO_2 that is emitted during the fires something to be added to Canada's Kyoto balance sheet? If we continue to keep *provincial* scores on GHG emissions, should the carbon from the fires be attributed by province? From time to time the federal government has stated that the carbon burden will not be born "unevenly by province." But is this a realistic promise? Will we see Quebec pay a carbon transfer payment to Alberta? If carbon is truly going to carry a price tag in a carbon-constrained world, what does this do to the economics of the oil sands? To what extent should people driving cars and those living in detached single-family dwellings be expected to internalize (i.e. pay for) the environmental implications of their lifestyle?

In order to come to grips with these difficult issues we need to examine more closely the actions that have brought us to the current situation.

Chapter 2

How Did We Get Into This Mess?

The message from the Toronto Conference was clear. The Earth's atmosphere is being changed at an unprecedented rate, primarily by humanity's ever-expanding energy consumption, and these changes represent a major threat to global health and security. Sound policies must be quickly developed and implemented to provide for the protection of the planet's atmosphere.

The Changing Atmosphere: Implications for Global Security,
Conference hosted by Environment Canada, Toronto, 1988

Mr. Dion likes to tell us the planet's fate is in our hands. Sorry! It's not. It's a big old world out there, and most of the six billion people in it are scrambling to use more energy, not less.... Despite our good intentions, we can't do anything about it.

Margaret Wente, "Carbon Cuts are Just a Fantasy,"
The Globe and Mail, 24 June 2008

Throwing a Tantrum in the Global Nursery

Human beings are a recent arrival in the context of Earth's four-billion year history. For most of our 70 million years we kept a low profile, certainly much less important to the health of the planet than, say, bacteria. But suddenly about 10,000 years ago, humans abandoned their quiet, unintrusive lifestyle based on hunting, fishing, and gathering, and moved into agriculture.

Why human society took this dramatic turn has never been satisfactorily explained; but, whatever the reason, it took on a momentum of its own. Innovation produced more food, which supported more people, who in turn needed more goods, especially food and water, to stay alive. Whenever an innovation increased productivity, especially in agriculture, population increased, until food shortages returned. Thomas Malthus (1766–1834) argued that, unless they practised self-restraint, humans were trapped in an inescapable cycle of growth and famine. Others, like Adam Smith (1723–1790), were more sanguine, believing that human ingenuity would always rescue humanity from its folly. In a sense, we are now facing this same debate again. We may find a solution to the current problem, but then discover that our "solution" creates a new set of problems. Bill Vanderburg identified the "solution/new problem" cycle as *The Labyrinth of Technology* (2005).

This observation may seem tangential to the problem of climate change in Canada, but the reality is that it is central. The evolutionary process of perpetual innovation, regardless of the consequences, is the essence of our occupation of the planet in these last 10,000 years. Since the beginning of our transition to agriculture, we have been changing local climates: we have drained lakes and wetlands, dammed rivers, cleared forests, ploughed up grasslands, flooded valleys—usually to produce more food and other goods. What is new about our current predicament is that now we are changing Earth's climate on a *global* scale. No part of the planet will remain untouched.

What is ironic in the Canadian context is that global warming will have the most immediate impact on Aboriginal people in northern Canada, including the Arctic, where many Inuit still practice the hunting/fishing lifestyle on which humanity depended, until innovation took over.

The big question is this: can we learn enough about the changes we made to the planet so that we can immediately slow down, and eventually halt, climate change?

A Very Brief History of Our Use of Energy

This is not the first time that we have faced a major fuel crisis and associated environmental problems. In fact, the switch to fossil fuels was prompted by exactly such as crisis. In the fifteenth century, wood—the main source of fuel and construction material—began to run out.

The first country to face shortages was Portugal. Portugal pioneered the concept and practice of maritime empire which transformed Europe starting in the fifteenth century. Wood was a necessity in shipbuilding, and as Portugal began to establish a global empire, it required many more ships. The domestic wood supply was quickly exhausted. By the seventeenth century, the Portuguese were building their ships throughout their empire, including Bahia, Brazil, and Goa, India. Globalization is not a recent phenomenon.

As Britain became a leader in the race for global dominance it ran into the same limiting factor: not enough wood. In addition to Portugal's priorities, Britain also produced a great deal of iron, for which charcoal was the essential fuel. The British landscape was eventually stripped of trees, and the same thing happened in France, Germany, and elsewhere. Initially, Britain plugged the wood shortage by importing timber from Russia, the Baltic, and North America. Economic logic eventually convinced the Royal Navy that it was cheaper to build the ships in New York and New England.

The offshore construction solution was fine for the Navy, but it did not solve the domestic fuel-for-industry problem. Manufacturers were obliged to consider coal as a possible substitute, even though it was considered to be a vastly inferior fuel to wood: coal produced large quantities of gases and impurities. It had been burned as a household space-heating fuel in Britain since Roman times, but only in small quantities. This was about to change.

In England, the ironmasters in the Weald and the Forest of Dean were adamant that coal would not meet their standards; other potential users were less fussy. Coal could be used in existing wood-burning stoves; this solved the "infrastructure transformation" problem. Households began burning coal for cooking and space-heating purposes. For some industries, the gaseous impurities were not a problem. Coal was introduced into activities such as the evaporation of salt, the brewing of beer, and the making of soap. Early on, it was small stuff. Yet in this "small stuff" lies the beginning of our contemporary huge, global problem of climate change.

Gradually coal became the fuel driving the Industrial Revolution, powering steam pumps and steam engines. Eventually, cleaner-burning coal was accepted by the ironmasters, enabling them to build engines more cheaply. Steam engines reduced the problem of flooding in coal mines. Trains and rail-lines introduced an era of cheap transportation. Steam boilers and iron ships replaced wind-powered wooden ships. Methane gas was drawn from the coal seams to provide "town gas"—a cheap way to light factories, streets, and finally homes. Factory workers were able to work longer hours, and people could work or read in the evenings at home. Eventually "shift work" was introduced, so that factories could run for 24-hour days. In 1834 in London, the first electric power was produced by burning coal. Using the language of systems theory, a powerful set of *positive feedbacks* was now in place, driving the rapid growth of both population and economy, and the demand for fossil fuels.

Throughout the nineteenth century, sources of power became more diversified with the first commercial oil well coming into production in Pennsylvania in 1859 and the first hydroelectric power station in Niagara Falls in 1886. Natural gas, another fossil fuel, became important in the twentieth century, as did nuclear power. Renewables (excluding large hydroelectric power) always remained a small part of the mix; today these stand at 2 percent of global primary energy. Large hydro accounts for another 2 percent. Today approximately 80 percent of the world's primary energy comes from fossil fuels, all of which emit greenhouse gases as a by-product of combustion. In Canada the current energy mix is as follows:

- fossil fuels, 70 percent
- hydroelectric power, 11 percent

- nuclear power, 11 percent
- biomass (wood), 6 percent
- other renewables, 1.5 percent

Even George W. Bush was moved to observe that "we are addicted to oil." Like other addictions, this one comes with a serious downside.

How long have we known that we had a problem?

Warnings Ignored

We saw how Joseph Fourier identified the heat-trapping capacity of carbon dioxide as early as 1826 when the carbonization of the global economy had scarcely begun. Svante Arrhenius (1859–1927), a Swedish chemist, provided a more explicit warning with an article and a book at the turn of the century, in which he calculated (without benefit of computer) that a doubling of CO_2 in the atmosphere would produce a temperature increase between 1.5°C and 5.5°C. This is exactly the range predicted today by an array of computer models, built by teams of scientists around the world. Under the business-as-usual scenario we are well on our way to doubling atmospheric CO_2.

These early warnings were not only ignored, they were drowned out by the opposite concern: that we would run out of fossil fuels and the global economy would be stranded. Today people worry about "peak oil." At the end of the nineteenth century the big concern was "the coal crisis." What would happen when there was no more coal?

The issue of CO_2 finally entered the global science agenda indirectly. In 1957, at an archaeological conference on radiocarbon dating, certain anomalies were noted in the carbon content of specimens, suggesting that atmospheric CO_2 must be increasing. A young physicist, Charles Keeling, went to Hawaii to make direct measurements in the atmosphere from a site on the Mauna Loa volcano. What he discovered provided a classic of field observation, showing incontrovertibly that CO_2 concentrations in the atmosphere were steadily increasing.

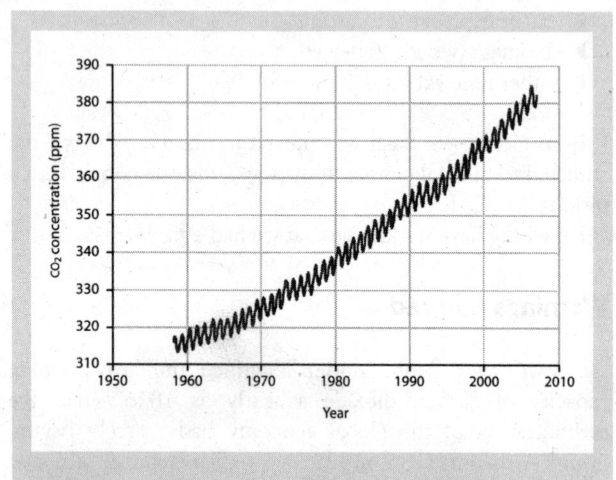

Figure 2-1 The Keeling Curve: Atmospheric CO_2 concentrations measured at Mauna Loa Observatory, 1958

Mauna Loa was selected because it was a very big mountain as far away as anywhere from industrial pollution. It would provide a good sample of a "well-mixed" atmosphere, with no local contaminants. Even within the initial few years of measurement, it was quite clear that CO_2 was steadily increasing in the atmosphere (the zigzag shape of the graph reflects seasonal changes in the concentrations). In the northern hemisphere summer season, tree growth absorbs (or sequesters) CO_2; in winter the lack of growth produces an "exhalation" of CO_2. As physicist Richard Peltier observed, "the graph showed the Earth breathing." These observations continue.

Subsequent research on ice cores extracted from the polar icefields has enabled scientists to trace this atmospheric CO_2 record back for more than a thousand years, showing that increased CO_2 coincides with the beginning of the Industrial Revolution. The Intergovernmental Panel on Climate Change

(IPCC) has subsequently filled in the details, from outcomes to impacts to possible solutions.

Even before the Rio Earth Summit in 1992, scientific consensus was growing that global warming was a big problem, requiring urgent attention. A milestone for focusing this concern was a conference sponsored in 1988 by Environment Canada. Hosted in Toronto, the results of "The Changing Atmosphere" were presented in an executive summary. This was a brief, readable report, available free to anyone. Subsequently, countless reports have been produced—by the United Nations Environment Programme and many non-governmental organizations, for example—for the politicians, business leaders, and the general public.

John Firor, the former director of the Advanced Study Program at the National Center for Atmospheric Research in Boulder, Colorado, has been a prolific public speaker on the subject of atmospheric change. He reports that the most frequently asked question from the audience was: "Tell us straight—are we in trouble or not?" In the conclusion to his 1990 book, *The Changing Atmosphere*, he wrote,

> In summary, the best efforts of the world's scientists foresee a rapid heating of the climate, and a vigorous search for reasons not to be concerned about this change has so far failed (96).

All this information was available over twenty years ago. So why is it taking so long for the message to sink in?

The Lure of "Liquid Gold"

There are many reasons why we have resisted accepting our role in actively warming the planet by chopping down trees and burning fossil fuels. As Al Gore put it, "climate change was an *inconvenient* truth." It is inconvenient because our response requires a fundamental rethink about how most of us live our lives. This may not include the Inuit in the Arctic, or the Khoisan in the Kalahari, or other hunters and gatherers around the world. But for most of Earth's six billion inhabitants, an adequate response implies significant change.

Canada is perhaps an extreme case. Our country is the most energy-intensive per capita in the world (in the previous chapter, we considered the reasons for this). Not only do we use a lot of energy, we also produce and export it. In addition, we are the number one external source of oil for the United States. As the price of oil rose in 2008, new oil and gas projects became viable; this ensured that Canada remained a key supplier. This is another important positive feedback that drives climate change. There is considerable pressure to develop the oil sands in Alberta and Saskatchewan, the offshore oil in the Hibernia field, and the increasingly "accessible" submarine Arctic oil and gas. There is even enthusiasm for developing natural gas from methyl hydrate resources in the ocean bed off British Columbia. Given the energy required for extraction, methyl hydrate is the natural-gas equivalent of the oil sands. Is there nothing we do not consider as a potential source of additional fossil fuels? Apparently not.

There are at least two good reasons why we might pause to reconsider Canada's relentless enthusiasm to continue unchecked along the fossil-fuel path. The first reason is the widespread understanding that climate change could have very nasty consequences worldwide, *and especially in Canada*. Second, these "alternate" energy sources require not only a lot of energy to extract, but also a lot of water. We will see in chapter five how the Prairies are already Canada's most water-stressed region. This fact is usually presented as an "environmental issue" that threatens habitat for wildlife. It is also an economic issue which threatens every aspect of the regional economy, especially agriculture.

Climate change is already shrinking glaciers that feed Prairie agriculture. Plans to "enhance oil recovery" by injecting water underground to force oil to the surface will require additional water. All this appears to present us with a regional recipe for self-destruction.

Reading a newspaper in Canada provides evidence of some kind of national schizophrenia. There are articles that report the dangers of climate change and the measures underway to begin to address the issue. In the same paper—sometimes *on the same page*—you can read articles about the latest plan "to develop additional oil and gas resources," that is, to emit even more GHGs. Until recently, the articles on oil and gas carried no mention of the associated climate change risk. This situation began to change in August 2008, especially once it seemed likely that the United States—now driven by new zeal to address the climate challenge—would impose tariffs on Canada's "dirty" (i.e. carbon-intensive) oil sands. We will come back to this issue later.

Climate Change Initiatives from Industry and Municipalities

Within the oil and gas industry, there are many concerned people and companies who understand that a fundamental shift is needed to reduce the risk of planetary turmoil under climate change. On the international scene, British Petroleum (BP) was the first oil and gas major to publicly announce, in 1997, that the current trajectory had to change. Within Canada, TransAlta was one of the first fossil fuel companies to develop activities to promote sustainable development. Elsewhere among Canadian energy providers, Hydro Québec and Enbridge have been some of the first movers.

Other sectors of the economy have also begun to act. The insurance industry has learned hard lessons from industry's neglect of the potential costs of pollution (asbestos, lead, toxic landfills). Insurers have been applying these lessons to the potential costs of climate change by linking it to the occurrence of extreme weather events, such as drought, flood, tornadoes, and heavy rainfall.

Municipalities were among the first to recognize the dangers because many extreme weather events took a major toll within their borders. As noted on page 19, as early as 1988 Toronto was the site of Environment Canada's international conference on "The Changing Atmosphere: Implications for Global Security."

> The message from the Toronto Conference was clear. The Earth's atmosphere is being changed at an unprecedented rate, primarily by humanity's ever-expanding energy consumption, and these changes represent a major threat to global health and security. Sound policies must be quickly developed and implemented to provide for the protection of the planet's atmosphere (Foreword to the Conference Statement).

Twenty years ago there was already a sense of urgency and a clear understanding of what, specifically, "climate change" meant. Even before the voluminous reports of the Intergovernmental Panel on Climate Change began to appear, experts argued that "the accelerating increase in concentrations of greenhouse gases in the atmosphere, if continued, will probably result in a rise in the mean surface temperature of the Earth of 1.5 to 4.5°C before the middle of the next century." Furthermore,

> global warming will accelerate the present sea-level rise. This will probably be in the order of 30cm but could possibly be as much as 1.5m by the middle of the next century. This could inundate low-lying coastal lands and islands, and reduce coastal water supplies by increased salt water intrusion.... The frequency of tropical cyclones may increase and storm tracks may change with consequent devastating impacts on coastal areas and islands by floods and storm surges (*The Changing Atmosphere* 3).

The message from the conference took only 30 pages to summarize (in both official languages) and could not have been clearer.

The International Centre for Local Environmental Initiatives (ICLEI) chose Toronto as its headquarters when it was founded in 1990. Canadian and American cities were among the founding members. At the time of the Kyoto Conference (which produced the Protocol in 1997), 145 local government organizations staged their own climate change event at nearby Nagoya. They too were right on message in accepting "the conclusion of the *IPCC Assessment Report* that stabilization of the global climate may require reductions in greenhouse gas emissions by more than 50 percent" (*The Final Nagoya Declaration*, quoted in White [2002]).

Eventually, in February 2005, the Kyoto Protocol came into force when more than 55 of the signatories to the Protocol, representing more than 55 percent of that group's GHG emissions, had formally ratified it through their legislative systems. In anticipation of this, the European Union established its own cap-and-trade Emissions Trading Scheme, beginning in January 2005. In North America (as of May 2009) all that we have in place is a patchwork of local initiatives and a great deal of voluntary activity such as the Chicago Climate Exchange. Europe is up and running, but the rest of the world has barely made it to the starting gate. This has not gone unnoticed.

Waiting for America: Eight Lost Years

From the beginning, climate change has been a highly divisive issue. This was clearly illustrated when George W. Bush became US president in 2000. President Bush promptly withdrew America from the Kyoto process, citing the need for developing countries to take on emissions caps in order that the American economy would not be disadvantaged. He remained unmovable on the issue right to the end of his second term.

Other parties in America have not been idle. The Chicago Climate Exchange was the first carbon trading floor in the world when it opened in 1993, supported by an impressive array of major corporations like American Electric Power, Dow Chemical, DuPont, and the Ford Motor Company. (See www.chicagoclimatex.com.) As mentioned above, many American municipalities were founding members of ICLEI. Finally the US began to take initiatives, both unilaterally (like California) and in regional groups such as the Western Climate Initiative and the Regional Greenhouse Gas Initiative in the northeast. At the same time over a dozen bills on climate change and energy efficiency are making their way through Congress.

Despite Canada's early support for the Kyoto Protocol (it was ratified by the federal government in 2002), action on the ground at the provincial and federal level has been glacial. Only BC, Alberta, and Quebec have passed legislation, although the other provinces and the federal government have issued innumerable "plans." (The details of these plans are considered in chapter seven.)

Where Are We Now? Finding a Position between Denial and Despair

As the evidence for climate change becomes more obvious, the number of climate change skeptics decreases. Unfortunately they are being replaced by a defeatist brand of pessimism, which propounds that it is *such* a big problem that a solution is impossible, or at least is not likely. There is also a chorus of would-be "geo-engineers" who believe that we can install some sort of protective sun-shield in the atmosphere, or seed the ocean with carbon-absorbing algae, or breed trees to sequester carbon *permanently* in the ground, or various other unlikely fantasies. No doubt we will be hearing more of these "with one bound, Jack was free" proposals, but we will not discuss them in detail here. However, the creeping pessimism could be a serious problem if it spreads.

Some of the pessimists (including environmental scientist James Lovelock) are in favour of "the nuclear solution." For example:

> Besides, what choice do we have as the existing nuclear plants come to the end of their life cycles? More power from coal, with its greenhouse gas emissions? Or from natural gas, which has become a very expensive fuel? Or from wind, which doesn't always blow? Or massive savings from conservation, which would require a dramatic change in our lifestyle? (*Toronto Star*, 22 June 2008)

Then there is the worry that even if "we" (in the richer countries) reduce emissions, it will make little difference to global warming because "they" (in the poorer countries) will increase their emissions more than we can cut ours.

> Mr. Dion likes to tell us the planet's fate is in our hands. Sorry! It's not. It's a big old world out there, and most of the six billion people in it are scrambling to use more energy, not less.... Despite our good

intentions, we can't do anything about it. (*The Globe and Mail*, 24 June 2008)

In this book I will try to lay out the situation in an objective, non-ideological fashion, leaning neither left nor right, not giving in to resignation to our inevitable, bitter fate, nor embracing nuclear power as the only solution. Nor will I throw in my lot with the "geo-engineers" advocating some technological fantasy.

There are a variety of possibilities out there to help us reduce our GHG emissions—and "theirs" as well. Some of these involve changes in lifestyle: less flying, less driving, smaller houses, and better insulation for some, perhaps. There are also some very important technological innovations that would help significantly. These include fuel cells, carbon-capture-and-storage, and a breakthrough in power storage. Humanity has been through many crises before—world wars, devastating plagues, adverse weather—and there is no reason to assume, *a priori*, that climate change will be our final one.

It is worth noting at this point that there was one very significant (and fairly recent) crisis for the evolution of human beings that, for the most part, has passed unnoticed. The Industrial Revolution drew us into the fossil fuel age. It was also a time of rapid urbanization. Yet, for a long time, mortality rates in cities exceeded birth rates, making cities dependent on a constant influx of rural people to keep them going. The lack of clean water and sanitation favoured many kinds of disease such as cholera, typhoid, typhus, plague, and other transmissible diseases. If you had stood back and assessed humanity's chances of making cities habitable for rich and poor alike you might have been pardoned for a certain amount of pessimism about how this was all going to end.

The problem was solved with the eventual realization that a solution had to involve the rich *and* the poor. Otherwise it would not work: the poor would continue to serve as hosts for diseases, and these diseases would strike the rich. You might live in the West End of London, relying on the prevalent westerly winds to blow the pollution away, over poor people living in the East End. But transmissible diseases are less easy to escape. Eventually, the solution to the urban health problem was something called "The Public Health Idea." This established that "health" should be treated as a public good, beginning with the low-cost provision of clean water and sanitation to everyone, including the poor.

This observation is also a clue as to how we might one day address the global warming problem. We have to find a solution for the poor as well as for the rich. Otherwise, there is no solution for anyone.

This is not the time to throw up our hands and say that anything "we" do will be nullified by "them." Nor is it the time to say that if "we" cap our GHG emissions then we will put a carbon duty on goods that we import from countries without caps. Canadians need to think now in global terms.

But first let's look in a little more detail at where the current trajectory on climate change may be taking Canadians.

Chapter 3

Likely Impacts on Canada and Canadians

The impacts of changing climate are already evident in every region in Canada.

Resource-dependent and Aboriginal communities are particularly vulnerable to climate changes. This vulnerability is magnified in the Arctic.

Canada is projected to continue to experience greater rates of warming than most other regions of the world throughout the present century.

Natural Resources Canada, *From Impacts to Adaptation: Canada in a Changing Climate 2007*

Climate change can introduce two major changes for electricity transmission and distribution. First, transmission losses will increase in higher temperatures and second, transmission infrastructure may be damaged more frequently by extreme weather events such as ice storms, tornadoes, and cyclones.

Environment Canada, *Climate Change and the Canadian Energy Sector*

Some Basic Terminology

How do we measure global warming?

The Earth's average surface temperature is measured as an annual average. This is based on thousands of readings taken at scientific stations on land and from ships, as well as other monitors at sea. It is averaged throughout the seasons at all of these locations. This process produces a very "conservative" indicator, in the sense that it responds to so many different signals that for it to *move*, in the same direction, over a number of years, cannot be the result of chance. Something, some process, has to be behind the change.

We now know in some detail how the extra heat-trapping effect of the six greenhouse gases works. Essentially these emissions have overbalanced the existing natural cycles of carbon dioxide, methane, and nitrous oxide. In other words, the sources of these gases have exceeded the *sinks* in which they reside at different parts of the cycle. As the *atmospheric sink* becomes overloaded, it traps more of the short-wave radiation that is reflected back from the earth's surface.

Scientists have built computer simulation models known as "General Circulation Models" to estimate the impact of the extra emissions on the global surface temperature, using various emissions "scenarios," of which the most basic is the "business-as-usual" scenario. Business-as-usual, as the name suggests, assumes that humans make no significant effort to change their ways, that the human population and the global economy continue to grow, and that the associated additional GHG emissions increase. The next step is to ask the following question: by what date, under a business-as-usual scenario, will atmospheric CO_2 reach a concentration that is double what it was before the Industrial Revolution? Finally we ask: when the doubled concentration occurs (e.g. in the year 2050) what will the globally averaged surface temperature be, compared with the pre-warming temperature?

Having established this "baseline calculation"—doubled CO_2 under the business-as-usual scenario—climate scientists then rerun the models under different GHG assumptions. This allows them to predict the likely impacts of various "mitigation strategies," such as assuming the adoption of best available technology for various sectors of the economy.

"Mitigation" is important in climate change terminology. Here the word refers to any efforts to reduce GHG accumulation in the atmosphere. Reductions may come from reduced GHG emissions or from increased uptake of atmospheric GHGs by another sink in the carbon cycle. For example, increased uptake of carbon dioxide can be achieved by planting trees or converting arable land to grassland. Both of these land-use activities "sequester" (or bury) carbon that would otherwise pass into the atmosphere. In fact, mitigation is a vast field that we will explore throughout this book.

In the late 1980s and early 1990s some of the people who were worried about climate change would speak *only* of mitigation as the best response to the threat. They did not want to encourage discussion of "adaptation" to climate change, in case that gave the public the impression that society *could* adapt. Soon, however, it became widely understood that some climate change was inevitable, whatever heroic efforts were made on the mitigation front. This new assumption was described as "the climate change commitment": there is no way around the fact that some changes will take place.

The earliest ideas on adaptation were pretty direct, even naive perhaps, like building bigger seawalls. Later, a more complex suite of responses evolved, such as improved preparedness for disasters like floods and hurricanes. Insurers invested time and energy in improving adherence to building codes (enhancing the integrity of the roof of a building in a storm, for example) and zoning regulations (keeping new buildings out of flood plains and vulnerable coastal zones).

Another piece of terminology is "carbon" as an adjective to refer to pretty much anything associated with climate change. For example, we have "carbon finance," "low-carbon technology," "carbon risk," the "carbon economy"; companies claim to be "carbon neutral" (meaning that their net GHG emissions are zero), while "carbon brokers" will source "carbon offsets," and so on.

The terminology arises from the Kyoto Protocol. Most importantly, those countries that pledged to reduce their GHG emissions below their 1990 baseline made a "Kyoto commitment." For example, Canada pledged to bring emissions to 6 percent below the baseline. This reduction is to be achieved in the "first Kyoto Commitment Period" which runs from 2008 to the end of 2012. Nothing has yet been agreed on what will happen after 2012—this uncertainty is known as the "post-2012 risk."

What are the most likely impacts from climate change? What do the global circulation models tell us is in store for Canada, for example? The rest of this chapter will provide an overview of these predicted impacts, and these impacts will be elaborated in regional case studies in chapter six.

Warming

The most predictable climate parameter is surface temperature, which is expected to generally increase throughout almost all of Canada. The one exception to the warming trend is coastal Labrador, which may actually cool for a number of years due to an increase in the number of melting icebergs in the Davis Strait and the Labrador Sea. There is evidence of such a change going back to the second half of the twentieth century. The greatest warming will occur in northern and Arctic Canada, as the ice and snow cover melts, thereby reflecting away less and less of the incoming sunlight.

More of the warming will occur in the winter months, rather than the summer. However, the warming that does occur in the summer will increase the number of very hot days. This poses a significant health risk through heat stress and smog, especially in urban areas.

After the north, the region most at risk to warming is the Prairies where more heat will increase the rate of "evapotranspiration" and hence reduce the moisture that is available to plants. The risk of drought will increase in the Prairies and also in the Okanagan Basin in central British Columbia.

Higher temperatures will increase electricity transmission losses, evaporation losses from hydroelectric dams, and probably lower the levels of the Great Lakes. Lower lake levels may force shipping companies to use smaller, shallower-draft vessels; however, this threat to the profitability of the system may be compensated by a longer shipping season. Warmer winter temperatures in the Arctic may reduce the efficiency of equipment used in the oil and gas industry. The warming impact will first reduce the usefulness of "ice roads," which are very significant in this region.

Impacts on animal habitat have already been observed. The most publicized so far is the reduction in ice floes which serve as platforms for polar bears hunting seals. The lower body weight of the bears when they come ashore to give birth has already led to reduced birth weight and higher mortality among pups. This in turn reduces food availability for the human hunters of the bears. Negative impacts on caribou have been observed, as on fish productivity as a result of the warming waters in north-flowing rivers.

Precipitation, Extreme Weather Events, Sea-Level Rise

The net impact of climate change on precipitation is more difficult to predict than the temperature changes; the reason for this is that precipitation is much more variable than temperature over space and time. Some areas may become wetter, and others drier. What is expected is an increase in the frequency of both heavy rainfall and droughts, more intense thunderstorms, hurricanes, tornadoes, and hail.

Overall, these changes increase the risk of mudslides, especially in the Rockies. More intense rainfall increases the risk of floods. We have perhaps already seen examples of these types of events with the heavy rainfall experienced in the Vancouver area in November 2006 and the downpour in North York, Toronto, in August 2005. In terms of major floods, which may be associated with climate change, we experienced the Saguenay flood in Quebec in 2006, and the Red River flood in Manitoba in 1997. (These specific events will be discussed later.)

Warmer spring weather may increase the occurrence of ice jams on major rivers like the Mackenzie. Hotter summers increase the risk of forest fires. Climate change may produce some counterintuitive effects, like the cooling of the Labrador coast mentioned above. It is possible that the near-record snowfall experienced in southern Ontario in the winter of

2007 to 2008 was partly due to the warmer winter keeping the water in Lake Huron and Georgian Bay more open than usual, making more water available for pick-up by the prevailing winter winds.

Some sea-level rise will occur, whatever happens to the Arctic and Antarctic ice. This is because the surface water of the oceans warms and expands, bringing both the danger of salt-water intrusion of aquifers and an increase in the height of storm surges. However, despite earlier uncertainty on the fate of the poles, the rate of melting and ice break-up—at both poles—now seems to be happening much faster than was first expected. In Canada, the places at most immediate risk to sea-level rise are the Arctic, Prince Edward Island, and Vancouver.

Air Quality

Higher temperatures are generally adverse for air quality given that they enhance the formation of ground-level smog from the hydrocarbons emitted by our fossil-fuel-burning activities, especially transportation. Smog formation, plus the direct stress on humans from higher temperatures, will threaten already vulnerable populations such as the elderly and asthmatics. One curious detail of climate change is that the warming trend will be more marked in nighttime temperatures than in the daytime. In fact, this difference is already observable in Toronto. This is a critical problem for people who are heat-stressed: the cooler nighttime temperatures potentially reduce the level of stress they experience over the daily range. When the nights stay warm, the mortality rate from heat stress goes up.

Heat stress is a serious problem in urban areas: temperatures are already elevated several degrees above the surrounding countryside due to the "heat island effect," whereby urban air is reheated by radiation from buildings and road surfaces, which absorbed heat over the course of the day. As 80 percent of the Canadian population now lives in cities, this could be a major health problem that will be exacerbated by both climate change and the aging of the population. Canada has not yet experienced a major heat stress event such as the one experienced in Chicago in 1995 (when 600 people died), or the European heat wave in 2003 (when over 40,000 people died).

Insect and Disease Vectors

Although people often complain about the severity of the Canadian winter, it does afford an important level of protection from diseases and their hosts that cannot tolerate low temperatures. We already saw how warmer winters enabled pine beetles in Western forests to spread, most recently in Alberta. The beetle lays its larvae in the bark of pine trees, especially lodgepole pine. As the bark dies, so does the tree. The dead wood is much more vulnerable to fire.

At least the pine beetle is indigenous to Canada. Forest managers have lived with it for decades. However, warmer winters are already allowing exotic pests like the West Nile Virus (which is carried by blackbirds) to survive. As its name suggests, this virus was first identified in Africa and made its way to the Middle East and the Eastern Mediterranean in the 1980s. It arrived in New York City in 1998, probably on a livestock-carrying ship. By 2004 it had spread right across Canada and the US.

These are only two examples of new disease threats that are amplified by the warmer temperatures associated with climate change. Other threats include Lyme disease and dengue fever.

Possible Benefits Associated with Climate Change

Surely there must be some potential benefits associated with warming up a cold country? There are, but they are few—and the benefits are far from assured.

Agriculture could benefit from a longer growing season in southern Ontario. Winter wheat, for example, could be encouraged; corn and soybeans could be planted further north. However, northward expansion of crops in the Prairies and southern Ontario will be constrained by poorer soils and probable lack of water.

One big predictable benefit will be the steady reduction in the demand for winter space-heating. There will however be a commensurate increase in the demand for summer space-cooling. Peak annual energy demand in Ontario, for example, already shifted from winter to summer in 2005.

Potential benefits may also accrue to those people, businesses, and institutions that learn to make the best of the "low-carbon economy." Manufacturers making vehicles and infrastructure that support a shift to public transport will no doubt benefit. Likewise, providers of "green buildings" and "low-emission automobiles" should benefit hugely.

People, Businesses, and Institutions under Pressure

Within Canada the people most at risk to climate change are the Aboriginal people of the north where the warming is happening most rapidly. For those who practise a traditional lifestyle based on hunting and fishing, dramatic change has already occurred. It is not at all clear where it will end, even if we begin aggressive mitigation efforts very soon. We have already seen how the warming effect will actually make oil and gas resources in the region more accessible. Even some Aboriginal people favour this form of resource development as a means to improve local standards of living. They criticize non-governmental organizations that argue against further fossil fuel development, accusing them of interference.

Another group whose livelihood may be undermined are Prairie farmers, who work in an uncertain environment even at the best of times. There are two principal new threats to their water supply: one is the loss of glacier-fed streams on which they depend for irrigation, and the other is the increased demand for water to improve yields from oil fields. The oil sands developers are also draining wetlands to allow open-cast mining. As in the Arctic, one form of development threatens to destroy an established way of life. Who will assess the trade-offs?

A third group are vulnerable people living in cities, facing the double threat of rising temperatures and deteriorating air quality. This group includes the elderly, the young, asthmatics, and others with respiratory problems. Poverty and isolation add to the risk. The deaths that occurred in the Chicago heat wave were all elderly people, living alone and mostly without air conditioning. Cities around the world, including Montreal and Toronto, have developed "heat response plans"; but these plans are reactive, rather than proactive. Unless the underlying issues of poverty and isolation can be tackled, the response plans can do very little to reduce the underlying risk.

Businesses, like individual people and families, are also under pressure from climate change. Farming and forestry are clearly at risk. A large agricultural corporation may be able to spread its risk by diversifying out of the most threatened regions. However, this is not an option for small family businesses, except to the extent that many farming families rely on off-farm employment to supplement their income. Clearly there is scope for developing new drought-resistant and disease-resistant strains of crops, or switching to crops that require less moisture.

Power producers, likewise, face a changing landscape. Although efforts are being made to reduce demand, most current initiatives will be overwhelmed by increased demand for space cooling in the summer. On the supply side there will be downward pressure from reduced water availability for hydroelectric power due to increased evaporation losses, as well as transmission losses in a warmer world. Thermal and nuclear power plants require water for cooling. Water supply becomes critical during hot weather (when power demand soars); this was evident during the European heat wave of 2003 when power plants in France and Italy were forced to shut down.

Transportation is another sector that is vulnerable in many ways, from lower lake levels affecting shipping to high winds and heavy rains interfering with highway traffic. Climate change creates pressure to shift from private cars to public transit and from trucking freight to shipping by rail. So far, little has been seen of this shift in Canada. Aviation too will be under pressure, as has already been seen in the European Union where plans are being developed to bring aviation into the European Union Emissions Trading Scheme. Ocean shipping is also on the list. In Canada, as noted above, the most immediately vulnerable transport links are "ice roads" in the north, winter roads built across frozen lakes or rivers. Already, the number of months for which these roads are usable is declining. In 2008 when ice roads started melting earlier than usual, the owners of the Yellowknife diamond mine had to shift from ground transport to aircraft for transporting remaining supplies, creating another positive feedback to increasing GHGs.

It is hard to think of any line of business that will be completely unaffected by climate change. The insurance industry was the first to respond to the changing landscape of risk. The industry realized that the historical records of insurance claims—on which the policy premiums are based—no longer reflected their exposure. Tourism—as a largely outdoor pursuit—will likewise have to make serious changes in order the stay in business. The days of Canadians flocking to Florida, the Caribbean, and Cancun during the cheap (that is, hurricane) season may well be coming to an end. As noted earlier, there will be opportunities in the low-carbon economy, and those companies that can turn a challenge into a profit will likely prosper.

Just as families and businesses are under pressure from climate change, so are all of our institutions throughout government and civil society. All levels of government should be involved in responding to these challenges. Generally, in Canada as elsewhere, local government has moved out ahead of the provinces and the federal government because they have to deal with the immediate problems on the ground, from floods to heat alerts. Extreme weather events reveal the cracks in society that might be less conspicuous in quieter times. Ongoing issues with homelessness, mental health, poverty, and isolation suddenly translate into large numbers of people at risk. Other municipal problems are more mundane, such as the increased cost of road repairs resulting from the more frequent freeze-thaw cycle experienced in a warmer winter.

For the bigger, longer term issues a commitment must be made by senior levels of government to invest in new, climate-proof infrastructure—physical, financial, and social. The contamination of a Walkerton, Ontario well with E. coli in 2002 was driven by unusually heavy rainfall which carried the contaminated cattle feces from a nearby farm to the well. The heavy rainfall can be viewed as a "climate change event," but the fatal impact on the people of Walkerton was a result of inadequate provincial surveillance of those responsible for testing and treating the water before distribution. The outcome could therefore be seen as a product of both provincial policy and climate change. Increasingly these various contributory factors will become entangled and produce unexpected results.

Every educational institution, from kindergarten to graduate school, needs to position itself to face the challenge, both in developing an appropriate curriculum and by greening its own operations. Some schools have been active in this area for many years; others are only just appraising their situation. Some non-governmental organizations may feel that they have been sounding the climate change alarm bell for many years, when few parties wished to hear the message. But even within the specialized *environmental* NGO community (ENGOs) there are some tough decisions that will need to be made. For example, should oil and gas development in the Arctic be encouraged, or opposed? Should carbon trading be supported, or should it be dismissed as merely "trading pollution"? Should ENGOs ally themselves with the greener aspects of the automotive industry? The list of difficult questions will only continue to grow.

The New Bottom Line

The ENGOs have been saying for years that the sooner we take action on climate change, the better for everyone. The skeptics dismissed such claims as being self-serving and alarmist. In October 2006 a report was issued by the British government which made the very same point as the ENGOs, in quite emphatic terms that even a skeptic would find hard to refute. It was crafted by a team of researchers under the direction of Sir Nicholas Stern, formerly chief economist at the World Bank.

The report (now available as a popular book, *The Global Deal*) was based on various scenarios relating to the possible severity of the impacts of climate change and various mitigative strategies. All produced the same conclusions:

- The probable costs (impacts) of inaction vastly outweigh the costs of mitigative action on climate change.
- The sooner we move, the lower these costs of mitigative action will be.

This was essentially the same conclusion that was drawn from Toronto's "The Changing Atmosphere" conference in 1988, over twenty years ago. Some might like to blame the US administration under George W. Bush for "eight lost years," but the fact is that we have all been guilty of wasting that time.

The implications of climate change for Canada are serious. It is not too late to move towards programs that will reduce the worst impacts and even produce a net benefit for society in some areas. However, this is not simply a domestic issue for domestic action. The broader implications of whatever action is taken also depend on the rest of the world, both for mitigation and for managing the impacts that are already inescapable.

The next chapter will assess Canada's position in the wider global context.

Chapter 4

Canada in the Global Context

Diseases currently prevalent in warmer climates will become increasingly greater threats in Canada as a result of greater incidence of disease and vectors in countries that are involved in trade and travel with Canada.

Many people will be forced to relocate internally within countries and internationally due to sea-level rise and growing water and food shortages in many countries, with implications for Canadian policies and activities related to aid, peace-keeping, and immigration.

Weather-related disasters, including drought, are projected to continue to increase in frequency and severity worldwide, resulting in increasing need for disaster relief and assistance from Canada, and losses for those Canadians with business and property abroad.

Natural Resources Canada. *From Impacts to Adaptation: Canada in a Changing Climate 2007*

The Dual Challenge of Mitigation and Adaptation

Until now the international focus for climate-change action has been on mitigation, on reaching targeted reductions in GHGs below the 1990 baseline. This focus has highlighted the contrast between those Kyoto signatories who accepted a target to reduce their emissions (the richer countries in the world) and those who did not (the poorer countries). Some of these poorer countries, like China, India, Brazil, and Indonesia, are major GHG emitters; little surprise that from the beginning an awkward question was posed: "*When* will these poorer, major emitters accept targeted reductions?"

The problem is even more complex. Because the global system is already committed to some degree of climate change, we also need to consider adaptation as part of the global challenge. Some aspects of adaptation have been introduced in the Canadian context. However, the problem of adaptation is more difficult in poorer countries because they have less adaptive capacity—less scientific capacity to predict extreme weather events, more economic reliance on weather-dependent agriculture, less social and physical infrastructure for disaster response, and so on.

The Kyoto Protocol says very little about adaptation. So in the absence of a global plan, what should we assume will happen?

We already know how some people in poorer countries adapt to prolonged environmental pressure (especially drought) and lack of economic opportunity. They migrate to countries with better prospects, legally when possible, illegally when not. This is already a worldwide phenomenon without the added pressure of climate change. The term "refugee" was initially applied to people fleeing from political threats. We have seen the recent terms "economic refugee" and "environmental refugee" used to describe those who may not necessarily be under political threat, but simply lack any realistic hope of economic improvement, or whose livelihood has been eroded by adverse environmental developments.

How will poor people adapt under climate change? Improved building codes and land-use zoning enforcement will make little difference to families who have no shelter.

The most likely answer is that more people will try to migrate to countries with better prospects, both legally and illegally. This expectation must be part of any realistic global approach to climate change.

A Unique Moment in Human History

We are facing a unique moment in human history. In previous disputes between the rich and the poor, the rich could dictate the terms of any settlement on a "take-it-or-leave-it" basis. The consequences of "no deal" were usually manageable for the rich, at least. Climate change is unlike World Trade Organization negotiations, or any other international forum. Now, failing to agree on a course of action is potentially catastrophic for *all* parties.

As long as enough people, businesses, and government bodies doubted the seriousness of climate change then this extraordinary fact could be set aside as irrelevant. Now that skepticism is seeping away, this is no longer possible.

We are confronted with a very stark situation. *Unless we can develop a cooperative solution, which leaves all parties better off, then we have no solution at all*. It matters, not just to Canadians, but to everyone else as well what happens in Chile, or China, or the Congo. This situation makes proposals for a "made in Canada" solution seem quite petty, not to say pointless.

Differentiated Targets: Why They Are Necessary

In the last fifteen years the global economy has undergone significant growth. Only this time, the growth has not been confined to the traditional post-war economic leaders like the US, Germany, and Japan; it has also taken place in what used to be called (at times optimistically) "developing countries." We have seen rapid growth in the world's two most populous countries, China and India. Both are not only populous but also are highly dependent on coal to drive their economies. In other words, their economies are highly carbon-intensive. Yet in per capita terms they emit less than 10 percent the emissions of North America. Their citizens are still relatively poor on average. In economic terms, a big "development gap" remains.

No plan for the global management of GHG emissions has a chance unless it recognizes this gap. Clearly targets for GHG management must differentiate between emissions from rich countries and emissions from poor ones. For example, against the 1990 baseline for the richer countries with Kyoto targets, poorer countries might be given a target date set at some time further into the future (e.g. 2030); this would become a ceiling from which they must come down. Alternatively, an absolute total of emissions, above what they currently emit, could be set as a ceiling. Or, a target could be set in terms of per capita emissions, in tonnes of carbon per person per year. Any mix of such targets could be set by mutual agreement. In the meantime the richer countries would pursue their Kyoto targets, and set further targets through the rest of this century, and beyond if need be. This is what is meant by "differentiated targets."

Another way of looking at the gap between rich and poor is to assess the amount of GHGs that have already been emitted in terms of the countries of origin. As GHGs are a product of the Industrial Revolution, the early emitters were countries in those first industrial regions: Britain, northwest Europe, North America. This factor is sometimes called the "historic burden" of GHG emissions. In terms of "which country emits what," it is worth noting that China with its 1.3 billion people has only just surpassed the United States (with its 300 million people) as the world's leading emitter of CO_2.

So any successful global management plan for GHGs should contain these two features:

- all countries accept some kind of target;
- the targets will reflect the "development gap" and produce differentiated targets according to current levels of wealth.

The Current Situation

The current situation lacks a global GHG management plan, and is not viable. Eventually emissions will reach a "dangerous level." Although this phrase is used frequently, no one can really put a number on it. And whatever that number is, it will vary from place to place. For small island states like the Maldive Islands, at serious risk from rising sea levels, we have probably already passed that point. For large, rich countries, the dangerous level could still be a year or two (but not ten years) away.

Among the richer countries, only the European Union currently has a GHG management plan in place. Beyond 2012 even the EU has agreed on nothing yet. (The "post-2012" risk is a big problem for industry as it is already well within their equipment renewal cycle.)

On the other hand, global poverty is still a huge problem. Approximately 40 percent of the world's people lack access to safe water; 60 percent lack access to safe sanitation. Using *currently available technology*, closing this poverty gap (assuming the political will to address the problem) would produce enough additional GHG emissions to put us *far* into the "danger zone" of an uncontrolled climate situation, sometimes called "the runaway greenhouse."

We saw above how many of the world's people are migrating to richer countries, in particular to Europe and North America. We can assume that these movements will continue as long as the pattern of opportunity remains broadly the same.

So the current global situation can be summarized as follows:

- We have a very small window of opportunity to reach a global agreement on GHG targets.
- The global burden of poverty is still immense; its reduction implies an increase in GHG emissions, *over and above* the business-as-usual scenario.
- We are already witnessing global movements of population towards regions of higher opportunity.

Canadian plans for GHG management cannot ignore these global realities. As one of the richest countries in the world, Canada is an important player at the GHG bargaining table. In addition to being rich, it is also one of the highest per capita emitters of GHGs. In these circumstances the pronouncements we have heard about "made in Canada" solutions for climate change are gravely unrealistic.

Why "Muddling Through" Is Not a Good Option

The persistence of poverty delays the demographic transition from high birth rates and high death rates to low birth rates and low death rates, from rapidly growing populations to stable populations. Countries that are currently rich have been through this transition. For the early industrial countries the transition took roughly one hundred years; for Singapore it took only fifteen. The longer we delay making the end of poverty a global priority the higher will be the final, stable population. This basic fact has been well understood for much longer than the climate change situation.

A greater number of people, even with many still living in poverty, translates into higher GHG emissions at a global level. These twin processes—getting rich, staying poor—create positive feedbacks of the kind we do not want.

It is finally becoming clear that these big global issues are intertwined and cannot be addressed—let alone solved—separately. The establishment of internationally agreed differentiated GHG reduction targets cannot ignore this fundamental fact.

The global dialogue on these issues will be very difficult, much more so than international trade negotiations; these after all are simply about quotas, tariffs, and subsidies. And time is running out. For small islands states in the Pacific, the Indian Ocean, and the Caribbean it probably has already run out.

A Glimmer of Hope

It was evident at an early stage of the Convention of the Parties to the UN Framework Convention on Climate Change that poorer countries were not going to accept GHG caps in the near future. It was also clear that to simply ignore them, leaving them to develop as "high-carbon economies," would negate any efforts on the part of richer countries to rein in their own emissions. The Clean Development Mechanism (CDM) was proposed as a cunning approach that would avoid some of the carbon intensity associated with the economic growth of poorer countries by transferring low-carbon technology from richer countries to poorer countries, paid for by the richer countries!

Finally there was something on the climate change bargaining table that would bring the poorer countries into the party. But the inevitable question arose: why should richer countries pay for this? What was in it for them (apart from the obvious answer of taking a step forward towards reducing the GHG risk to the planet)? This was where the inventors of the CDM took the plan even further.

In return for funding this transfer of clean technology to poorer countries, the investors (based in the richer countries) could earn credits towards the GHG emission reduction targets their countries had accepted under the Kyoto Protocol.

Immediately, this highly original proposal was branded as a "sell-out," and worse, by some environmentally concerned people and parties. It allowed the richer, capped countries to transfer some of their GHG reduction obligation to poorer countries. In other words—using some fairly twisted logic—they were "exporting their pollution to the poor"! There were some difficult days ahead before the CDM was up and running, and its application does require, at times, some extremely tortuous reasoning. It has been agreed—at least in the EU—that CDM credits may account for only a "minor" component of the GHG reduction effort in each capped country. After months of haggling over how big "minor" might be, most EU countries settled on a maximum percentage between 5 and 10 percent of the total reduction.

CDM credits now constitute an important component of the growing carbon economy. China was the first of the potential "host" countries to develop a CDM project development capacity, with liaison committees established in every single province. India and Brazil were not far behind. More detail on the CDM, and individual case studies, will be provided in chapters five and seven. TransAlta was one of the first Canadian companies to "source" CDM credits.

Implications of the Global Context for Canada

In this chapter we have reviewed some of the global issues that shape the development of climate change policies in Canada. As we have seen, there is no "made in Canada" solution to climate change. All countries will sink or swim, together. For Canada's mitigation efforts to have any impact, all countries must accept a GHG cap. This will be uphill work on the diplomatic front and it will continue for decades. It is not a matter of just signing agreements with like-minded countries and investing in clean technology, although that will be part of the process. The new reality is that climate change obliges all countries to work closely together to reduce GHGs wherever possible.

The next chapter returns to a Canadian focus to examine in more detail the technical and political implications of GHG mitigation.

Chapter 5

Some Technical and Political Options for Mitigation

Science tells us that it may be technically feasible (although exceedingly expensive) to capture 90 percent of the carbon emitted by a new coal-fired generator, but just 10 percent of the greenhouse gases associated with the oil sands.

Gerald Butts, "Carbon capture technology no silver bullet for tar sands—only a small proportion of greenhouses gas could be sequestered," *Toronto Star*, 26 February 2009

Four days after Ontario announced its latest wind contracts, Canada's wind energy sector was stunned when the federal government brought down a budget that failed to renew a crucial program subsidy.

Shawn McCarthy, *The Globe and Mail*, 17 February 2009

Fuel Switching

Each source of energy is associated with some greenhouse gas emissions. Even for nuclear power and renewable sources of energy, some GHGs are emitted during the manufacturing, maintenance, and decommissioning/replacement phases. However, this contribution is very small compared with emissions from the combustion of fossil fuels. Within the fossil fuels, lignite (brown coal) is the most GHG-intensive, while natural gas is the least. One option, then, is to switch to a less carbon-intensive form of energy. For example, a big part of Ontario's plan to reduce GHG emissions is to close its major coal-fired power stations and replace them with natural gas. A similar switch was made by the UK in the 1980s.

In the longer term (30 years?), fossil fuels will gradually be phased out of the energy mix. Renewable sources of energy are clearly the most attractive alternative, especially wind, solar, and hydro power. Offshore wind farms generally provide stronger and more consistent power than onshore farms and usually face fewer planning hurdles, although they are more expensive to install and maintain. Tidal power and wave power are also making rapid advances. There are some critics of renewable energy who worry about "intermittency issues," such as the wind dropping and the grid system being left without power. However, this variability can be handled by switching to another source—renewable or not—given that wind does not drop everywhere at the same time. Considerable research into this issue has found that it is not a problem if wind power provides a maximum of 30 percent of the grid's power. Spain, Sweden, Greece, and Germany are already approaching this level. (It is worth having a look at the UK Energy Research Centre on this issue.)

For wind turbines that are off-grid, local storage using batteries is possible, although battery efficiency needs be greatly improved before this becomes an attractive option. Wind power, with suitable storage options, could be very important for Arctic communities that are currently dependent on highly carbon-intensive diesel generators.

Biomass is also considered to be renewable as long as supply is assured. Agricultural waste and wood waste provide a substantial proportion of the energy mix in Sweden. At the moment the most controversial contribution from biomass is ethanol derived from corn to add to the gasoline mix. Critics claim that ethanol produces as much CO_2 as gasoline when the production process is taken into account. It also needs a good deal of water to produce it.

The definition of what is "renewable" is hotly contested. When the UK introduced its "renewable energy requirement" it included the combustion of landfill gas (methane) as eligible, even though the average lifetime of a landfill gas plant is only 30 years. However, the requirements did not accept nuclear power, despite demands from the nuclear lobby. Likewise for the Clean Development Mechanism, large-scale hydroelectric power is excluded from the "clean" category because of the environmental impacts and displacement of people. But what is considered "large-scale" for hydroelectric power is also disputed.

Critics of nuclear power as the "obvious alternative" to fossil fuels list four potential problems. First, there is no accepted means for the long-term storage of high level waste because no community wants to live in the vicinity of a storage site—a clear case of the "Not In My Backyard" syndrome (NIMBY). Second, nuclear plants are very expensive to construct and seem to be prone to dramatic cost overruns, at least in the Canadian experience. Third, the decommissioning costs are difficult to estimate given lack of experience and the unresolved waste disposal issue. And finally, critics cite security issues, both operational (Chernobyl, Three Mile Island) and in terms of potential terrorist activity. Despite these objections the combination of climate change and the insecurity of supply of oil and natural gas has brought nuclear power back on the agenda for fuel-switching.

Doubtless, there will be an interesting evolution of fuel-switching options as research into clean technology advances. For example, in Toronto, the use of a deep lake water cooling system is finally in place in downtown office buildings, after twenty-five years of debate.

Demand Management and Energy Efficiency

A better option than simply switching the source of fuel is to use less energy. "Demand management" really means "demand reduction." There are many innovative new schemes, many of which will come online soon.

A new technology known as "smart meters" is designed to influence consumers' behaviour. Many Europeans would be surprised to learn that throughout much of Canada, electricity costs the same to households 24 hours per day; in Europe the charges vary for peak and off-peak times. A "smart" meter records usage by time of day. Peak energy costs more to produce as it must be generated intermittently to supplement the base load. Energy planners hope to "shave the peak" by encouraging consumers to schedule their usage around it. This strategy reduces both overall cost and GHG emissions.

Another kind of smart meter is being introduced in Canada that automatically changes room thermostats at the peak, raising the temperature setting slightly in the summer and reducing it in winter. This too should reduce overall GHG emissions.

There is also much potential in new urban infrastructure. The Green Building movement encourages an integrated approach to new buildings and retrofits to reduce heating and cooling demand through improved insulation. The design includes water-efficient appliances, reducing the energy needed to treat the saved water, and hence reducing GHGs.

Intensification of residential urban land-use can be significantly improved by fitting new houses into existing neighbourhoods; this reduces demand for expanded services (energy, water, waste collection, etc). Public transit also becomes more viable, in addition to more walking and cycling, and less driving. Most of Canada's urban infrastructure was built when energy was cheap (compared to today, and compared to Europe), so there is plenty of opportunity to make our systems more efficient.

In general "energy efficiency" meets the same demand for energy, using the same fuel source, but with better delivery and smarter usage. Simply taking a critical look at how we produce energy reveals many reduction opportunities. For example, the traditional thermal power plant lets *two-thirds* of the energy it produces be vented to the atmosphere as waste heat. New plants are designed to use that waste heat, a system known as "combined heat and power."

Similarly, water-efficient toilets use only one fifth of the amount of water compared with the traditional 26-litre flush. Again this shift saves all the energy associated with purifying and treating the water used in the flush, plus conveying and treating the waste water.

Carbon Sequestration

A second-best solution to the excess CO_2 problem (if efficiency improvement, demand reduction, and/or fuel-switching opportunities do not go far enough), is to trap more CO_2 underground. This would prevent it from reaching the atmosphere. Forestry, agriculture, and industry all provide opportunities for what is known as "carbon sequestration."

Growing trees absorb carbon dioxide, therefore planting new trees will reduce atmospheric CO_2. However, when they die they release that CO_2. Hence there is a limit to the carbon reduction effect from planting trees. In farming when arable land is converted to pasture, more CO_2 is trapped in the soil, or, even if the land remains arable it can be ploughed using "minimum till," which releases less carbon than traditional ploughing methods. All of these approaches are being practised and some are receiving credits under the Clean Development Mechanism.

The great hope for the future is the application of "carbon capture and storage" technology to fossil fuel–fired power stations, which are major emitters of CO_2. A pilot project is under way in Saskatchewan (as well as at several other locations around the world). An even more efficient technology is to burn only "clean coal," which has been prepared in such a way that GHG emissions are reduced. As noted in one of the introductory quotations, carbon capture and storage will benefit new coal-fired utilities (at a cost), but will do very little to reduce the carbon burden of the oil sands development.

Methane Capture

Major sources of increased methane in the atmosphere include leaking gas pipelines, methane escaping from coal beds, methane released by landfills, paddy fields, biomass-burning, and methane produced directly by ruminants and escaping from animal manure. A range of interesting projects are now in place or under development to address these emission sources, such as improving the maintenance of gas pipelines. For many of the other sources, the methane can be captured and burned on site to produce power that would otherwise have been wasted. An interesting note is that there is enough methane available in a large-scale sewage treatment plant to power the entire plant; a plant just like this was constructed in Los Angeles in 1998.

Other Greenhouse Gases

Another important GHG is nitrous oxide (N_2O), emitted from both industrial and agricultural sources as well as biomass-burning. The major source of nitrous oxide is fertilizer.

There are three other important greenhouse gases: these are much smaller in volume but very long-lived in the atmosphere, so for this reason they have a high global-warming potential. Hydrofluorocarbons (HFCs) were introduced as a replacement for ozone-depleting chlorofluorocarbons and are used in industry and as a refrigerant. Perfluorocarbons (PFCs) are used in the aluminium industry, in electrical appliances, in fire-fighting, and as solvents. Another compound, sulphur hexafluoride (SF_6), is used in electronic and electrical industries and in insulation. Because of their very high global-warming potential these four industrial gases have become key in the Clean Development Mechanism system. They are easy to locate (mostly in factories and associated infrastructure) and earn many GHG reduction points.

Refitting a factory is certainly a much easier approach to GHG reduction than capturing and burning methane from pig manure.

Political Options for GHG Mitigation

The political options are even broader than the technical options, especially in a country like Canada with its vast territory and complex, often combative federal-provincial structure. However, we must ask: "Is it really any more complicated than the European Union which has already begun to establish a credible response to the climate change challenge?"

The general objective is to set up an *incentive structure* that will encourage GHG reduction among businesses, institutions and households. That structure could take on several different forms of which only the main types will be discussed below.

The EU Emissions Trading Scheme is the biggest example of a cap-and-trade system, under which each country in the EU was given a cap on its CO_2 emissions for the "First Compliance Period" (January 2005 to December 2007). Each country then set up a National Allocation Plan which capped every installation with a power production capacity of 20 megawatts per annum or more. The main burden of reduction was placed on the power producers, on the grounds that they were the least exposed to (uncapped) foreign competitors, and so could pass any increased costs on to their customers. In mitigation parlance this was an "upstream" approach, targeting "large stationary emitters." Eventually some of the reduction costs would reach their customers "downstream"—businesses, institutions, and households.

The heart of the incentive structure was the distribution of credits, or allowances, which emitters could hold, or sell into the market. At the end of the Compliance Period the emitters had to return to the Board a number of credits equal to the GHGs they had emitted. If they over-emitted they had to buy additional credits or pay a fine. If they under-emitted they could sell their surplus credits. In the subsequent Compliance Period the caps would be set even lower, and so on, until the Kyoto targets were met. These caps are "absolute" targets which must be met, or a penalty paid, with the missed target added to the next period's requirement for reduction.

It may sound novel and complicated. However, among the various options it is the only one most likely to meet targets on time, in a predictable manner, and at the lowest possible cost for the system as a whole. This approach was recently introduced in British Columbia.

Another approach is to "tax carbon": in other words, the idea here is to tax all fuels according to the amount of CO_2 they produce, whether by a business, an institution, or a household/individual. Each user "pays at the pump" as well as on the household energy or business energy bill. This is a "downstream" approach. It is impossible to say what size of tax will produce what quantity of GHG reduction given that the outcome depends on decisions made by millions of consumers. Over time it might encourage more carbon-conscious behaviour, or it might not. It is also politically very risky: few people like taxes. Quebec and BC have already adopted such a tax. The federal Liberal party (in Opposition) proposed introducing a carbon tax during the November 2008 election campaign, but as history records, they were defeated.

Technically, the impact on consumer behaviour of a carbon tax—which artificially raises the price of a commodity—depends on what economists call the "elasticity of demand" for the products that use fuel. If demand is "elastic" then a tax will reduce demand. As we noted above, however, nobody knows by how much. But if demand is "inelastic" then behaviour will change very little. As the National Round Table on the Environment and Economy pointed out: "There is no consensus on the price elasticity of gasoline" (1999, 68).

What do we know about demand elasticity for gasoline? We have two interesting anecdotal pieces of evidence. In Canada, when oil soared to $150 per barrel in the summer of 2008, the price at the pump jumped about 30 percent. Vehicle use in kilometres travelled dropped only about 3 percent. Not much elasticity there.

The second anecdote comes from the UK in the mid-1990s, when Chancellor of the Exchequer Gordon Brown boldly announced that gasoline taxes would be raised in order to reduce traffic congestion. The increase would be 6 percent the first year, 6 percent the next, and so on, until the desired effect was achieved. Even more boldly, he named his new tax an "escalator tax," to reflect its inexorably ascending nature. No one could accuse him of lacking political will or clarity of purpose. Alas, the British motorist did not share Mr. Brown's enthusiasm. The first year saw widespread protests. The second year saw gas stations attacked by outraged motorists; the army had to be called in to restore order. Mr. Brown turned his escalator off. A similar initiative in Belgium produced the same result.

Taken together, these two anecdotes suggest that applying a carbon tax to fuel (at least, to gasoline) will not significantly reduce GHG emissions. Consumer reaction and political caution will stop such a program dead in its tracks long before climate change is brought under control.

Lastly, among the main policy options currently on the table in Canada is the "intensity approach." This approach sets targets by industry for the amount of GHGs emitted per unit produced, per kilometre driven, per tonne transported, and so on. There is no overall cap, but as the intensity per unit produced is reduced then total, or absolute, emissions *may* fall. Or they may not, depending on how the economy grows and changes. Intensity target reductions are certain to produce an absolute GHG reduction only if the economy (or a particular sector) is shrinking. A GHG reduction scheme based on intensity targets has been introduced in Alberta, and was proposed by the Conservative Party forming the minority federal government. Given the huge growth planned for the carbon-intensive oil sands it is hard to see how such a scheme will produce an absolute decline in GHG emissions.

Many Canadian cities have been actively engaged in GHG reduction under the auspices of the International Council for Local Environmental Initiatives (ICLEI). An incentive to take the initiative (even though senior levels of government held back) was provided by the proximity of urban governments to climate change-related impacts on the ground—from flash floods, and an accelerated freeze-thaw cycle, to heat alerts and smog. It would be useful if some financial carrots could be developed to accompany these sticks. There is no reason why municipalities could not be brought into an emissions trading scheme and earn credits.

Municipalities enjoy the advantage of learning from the successes of others. In fact it was on this very basis that ICLEI was set up. Although cities around the world vary by climate, topography, economic and social composition, and political power, they all share the similar responsibilities to their citizens: education, health, transportation, protection from the elements and from crime, and so on.

By comparison, federal, provincial, and territorial governments have been slow to take up the challenge. The actions listed immediately above are the only ones to date, although countless plans have been produced (and will continue to be produced). Senior levels of government must act in a coherent way. To the extent that they do not, they impose an unnecessary burden on businesses and households—as taxpayers and as consumers—who will pay a heavy price in the long term. These prices include adverse physical impacts from climate change, lost business opportunities, and the increased costs of responding to the problem later, rather than sooner.

It is logical that a coherent and cooperative federal/provincial framework should be put in place before the chosen incentive structure (such as cap-and-trade) is linked to similar efforts in the United States and the emerging global carbon economy. Ideally this would bring more countries into the Kyoto tent and make better use of the Clean Development Mechanism to transfer technology from innovators to the rest of the global economy.

Manifestly this is not what is happening. The federal government has yet to move out of its planning phase, while several provinces have adopted observer status or joined various US state-level regional groupings. These include the Regional Greenhouse Gas Initiative (northeastern states), the Midwestern Accord, and the Western Climate Initiative. British Columbia has signed an international agreement to trade into the EU Emissions Trading Scheme. As far as the actions of senior governments are concerned, Canada is all over the map.

Next we will take a closer look at what the likely impacts of climate change will be in Canada, using illustrations from recent events with both regional and sectoral implications.

Chapter 6

Impacts on Canadian Resources, Livelihoods, and Quality of Life

Earlier this month, British Petroleum pledged to spend as much as $1.2 billion on searching for oil in three areas in the Beaufort Sea, the largest exploration commitment in Canada's history.

The Globe and Mail, 30 June 2008

Boil-water advisory hits two million in Vancouver—the largest single urban advisory ever made in Canada. Residents warned not to use tap water for drinking, brushing teeth, washing fruits and vegetables, after heavy rainfall triggers mudslides across region.

The Globe and Mail, 17 November 2006

Introduction: It's Mostly about Ice and Water

Eighty percent of Canadians live in cities, and those of us who do probably spend little time considering our debt to ice and snow—and especially ice.

Yet ice is greatly important as a form of free water storage; rivers draw on melting ice every spring. Even in southern Ontario, the spring snowmelt has a major influence over stream flow. For the rivers flowing east and west from the Rocky Mountains it is a vital part of the supply, much more dependable than sporadic rainfall. In Canada, as in many parts of the world such as central Asia, the glaciers and the snowfields are essential for maintaining this flow. It is not clear how we will function when the ice has gone.

For Arctic people, dependent on a traditional lifestyle, ice supports their entire livelihood. On this, and other occasions, you cannot help but be struck by the irony of decisions that are made—to extract oil and gas, for example—in the very same region where the most immediate negative impacts are going to occur.

The disappearance of snowfields and glaciers is a slow-moving event on the human timescale. It is only because glaciologists monitor glaciers every year that we can be certain that most of them are in retreat. This chapter considers first some of the slow-moving changes taking place, and then some of the extreme weather events that can occur very quickly, sometimes in just a matter of minutes.

No one can be entirely sure exactly what meteorological trends can be ascribed to climate change. Meteorologists know that climate change has the potential to shift major storm tracks, but they have not established that it generated the ice storm of 1998 which struck Montreal and the Ottawa Valley (which we will consider below). However, in Europe it is clear that the weakening of the winter high pressure system over northern Russia has permitted westerly storms to strike northern Europe with much greater severity. These storms are also penetrating further into the Baltic Sea than they have done previously. So it makes sense that we should be prepared for some large scale changes in weather patterns in Canada as well.

Rather than speculating about what form these larger changes might take, the chapter will focus on what we do know from the most recent scientific research.

The North

The Inuit, Indian, and Métis people who make up the varied aboriginal societies of northern Canada are all deeply attached to the land and living resources. Across the north these societies continue to depend on the land for their cultural and physical survival and

> well-being. Hunting, fishing, and trapping are not remnants of the past; on the contrary, food production and economic activity based on the land are very much part of modern aboriginal societies. (Mary Simon, Canada's first Ambassador for Circumpolar Affairs, 2003)

It is in Canada's far north that the effects of climate change first became evident. And it is here that greatest overall warming will continue to take place.

The first surfaces to be affected were the sea ice and the permafrost. Permafrost is land that remains below 0°C over two consecutive summers. Sea ice, as the name suggests, is ice that has formed over the sea, expanding and contracting with the seasons. However, over the last twenty years, this ice has slowly retreated, opening up new gaps between the remaining islands of land-based ice. Scientists estimate that Arctic sea ice has lost 40 percent of its mass over the last twenty years. We saw earlier how this has reduced polar bears' access to seals. It has also reduced the availability of polar bear meat to aboriginal hunters.

The connection between sea ice and the Inuit–polar bear–seal food chain is just one example of the many irreversible changes taking place in the Arctic environment. Other changes include the increased susceptibility of the caribou to disease and the warming (and possible destruction) of fish habitat.

The further irony of the disappearance of the sea-ice is that it makes undersea oil and gas reserves even more accessible for exploitation.

> Earlier this month (June 2008), British Petroleum pledged to spend as much as $1.2 billion on searching for oil in three areas in the Beaufort Sea, the largest exploration commitment in Canada's history. Imperial Oil also plans exploration work, and is pursuing a $16.2 billion gas pipeline that would bring Arctic gas to the North American market via NWT and the Yukon. (*The Globe and Mail*, 30 June 2008)

This is another example of the kind of positive feedback that will accelerate global warming. The implications of this trend are critical not only for the aboriginal people, but for everyone in Canada, as the disappearing sea-ice exposes Canada's claim to sovereignty.

And there are further implications. The disappearing sea ice exposes Canada's claim to Arctic sovereignty. Oil and gas are becoming more accessible, and the fabled Northwest Passage—sought by European explorers for 500 years—could soon become a reality. The US has never recognized Canada's sovereignty over the channel; this was the case long before there was any issue of commercial importance. Denmark, Russia, and Norway also lay claim to parts of the Arctic that are being opened up by global warming.

Water Levels in the Great Lakes

The net effect of climate change on water levels in the Great Lakes is still not known. Inflows are determined by rainfall, snowfall, snowmelt, runoff, and inflows from rivers; outflows are determined by evaporation, outflows from rivers, and diversions and consumptive use by humans. Two key variables are precipitation and evaporation. Generally, for the Great Lakes–St. Lawrence watershed it is expected that overall precipitation will increase. Another prediction is that evaporation will increase in a warmer world. So far, the warming in the region has been relatively small, compared with the Prairies and the Arctic. So we do not yet know what the net result of the precipitation/evaporation changes will be.

Direct observation of lake levels (taking Lake Huron as an indicator) shows a strong though irregular cycle, including variations within a two-metre range. From the late 1990s a marked (though not unprecedented) fall in the lake level has been observed. But what we don't know yet is whether this is part of the cycle of natural variability or whether it is the beginning of a trend driven by climate change.

A long-term downward trend in lake levels would clearly be a big problem—shorelines would recede and wetlands dry out. A decrease of water in the lakes would mean a reduction in hydroelectric power and shallower drafts for shipping. Residents and businesses here would find that docks no longer sit by the water and large expanses of polluted sediments are exposed.

Of course the consequences for wildlife, fish, and the native economy could be considerable.

Drought in the Prairies

The future of lake levels depends on the evolving balance between increased precipitation and increased evaporation. Similarly, the future of farming in the Prairies depends on the evolving balance between precipitation, evapotranspiration, and soil moisture. ("Evapotranspiration" is the transfer of water from the ground by evaporation, and the passage of water through plant matter, into the air.)

Certain factors would suggest we approach the future with caution. The Prairies produce 80 percent of Canada's agricultural output. The bulk of these crops are cereals, which are heavily dependent on irrigation. The loss of spring stream flow with the retreat of glaciers means that less water will be available, unless this is compensated for by greater precipitation. Even so, it is not just the *total* precipitation that matters. Crops rely on water from stream flow in the spring, during plant maturation; however, an increase in precipitation would most likely occur as heavy convective rainfall later in the season. This heavier rainfall closer to harvest time would actually reduce crop yield.

The Prairies are already subject to periodic droughts that can be severe enough to reduce harvests by 70 percent. Given the location of the Prairies in the interior of the continent, combined with the higher temperatures expected under climate change, there may be a decrease in total rainfall. At the same time, rainfall events may be more extreme. Generally this increases the risk of crop failure. Shortages will have subsequent effects in terms of food available for livestock (and livestock will likewise be stressed by water shortages).

That said, and despite the spate of serious droughts in recent years, there is as yet no clear trend towards more adverse conditions in the Prairies. Just as with lake levels, we can't yet say if the recent bad patch constitutes a trend. Indeed, the summer of 2008 was the wettest on record for many parts of Canada, including the Prairies.

It is worth noting that there is already intense competition for water in the Prairies between farming, urban demand, and oil and gas extraction. The expansion of the oil sands in conjunction with climate change will make this complex problem of resource allocation even more challenging.

The Okanagan Valley, British Columbia

The Okanagan Valley might have been designed as a laboratory for observing the impacts of climate change in Canada. It has the lowest per capita availability of water in Canada, yet it is a highly productive agricultural region—renowned for its vineyards, apples, apricots, cherries, nectarines, peaches, pears, and plums. It is the northernmost producing zone for most of these crops, and they are all entirely dependent on irrigation.

Another key feature is that the rivers in this area are themselves dependent on snowmelt for 50 to 80 percent of their water. Predictions for weather patterns in this region under climate change include warmer, wetter winters and warmer, drier summers.

Here is how a recent report summarized the situation:

> The semi-arid climate of the Okanagan Valley is ideally suited to grapes and tree fruit production which, along with tourism, are the mainstays of the economy. In most areas, agriculture is entirely dependent on irrigation supplied by annual runoff from the accumulation of winter snow in the surrounding mountains. (Cohen and Welbourn 2004)

There are other demands for water, including hydroelectric power, fish habitat, a growing population, and the tourism sector. Projections suggest that as a result of climate change, less water will be available in the summer months when irrigation demand is highest.

So the pressure on water availability comes from both sides. Other things being equal, demand for irrigation water will increase in response to higher evaporation rates. At the same time, less water will be available given the reduction of glaciers, decreased accumulated winter snowfall, and higher rates of evaporation from reservoirs. The big question for the Okanagan Valley is this: how will all these factors net out? What will be the final outcome for the Okanagan Valley? Will the anticipated increase in winter rainfall be sufficient to fill up the reservoirs and replace depleted glaciers and snowfields? Will that work for maybe ten years as the glaciers and snow gradually disappear? And after that, what will happen? Nobody can say with any real certainty.

But the recent past does provide a warning:

> Low snowpack in the 2002–2003 winter followed by record-breaking hot dry summer 2003 conditions created the worst drought the Okanagan has experienced in recent memory.... Record high temperatures and very low stream flows threatened fish and municipal and agricultural water supplies. Losses due to forest fires totalled hundreds of millions. These events may foreshadow future threats to the region under a longer growing season. (Cohen and Welbourn 2004)

Taking a "wait and see" approach seems unwise for many reasons. First, it makes sense to assess current water management practices to determine if they are the best available. We can also look at how other people have adapted in even more water-stressed regions around the world, such as the Murray-Darling Basin in Australia.

Clearly current water management practices in the Okanagan Valley are inadequate. Consider that water is still being measured by the "acre-foot": the amount of water it takes to flood an acre of land to the depth of one foot, the traditional method of "flood irrigation." (One acre-foot equals 1,233 cubic metres.) In the looming reality of water scarcity, many farmers have replaced wasteful flood irrigation with centre-pivot distribution, or, better still, "drip irrigation," whereby water is carried by a fine network of pipes and dripped on the plant roots. This is very well-suited for the tree crops grown in the Okanagan. But not all farmers in the region are on this system yet, partly because not all the water supply (agricultural or domestic) is even metered! With 100 percent metering *and* prices that reflect full-cost recovery, demand will probably fall, as it has elsewhere in the world.

In parts of Australia's Murray-Darling Basin, farmers are allocated water rights, which they can use, sell, or bank. This encourages saving water for the most valuable crops, and using it efficiently. Livestock producers, for example, may decide not to grow their own hay, but instead to sell their water rights and import the hay they need from another region.

Measures like these will surely need to be adopted in the Okanagan Valley as water becomes increasingly scarce.

Vancouver's Water Supply, November 2006

In contrast to the Prairies and the Okanagan Valley, it might seem unlikely that Vancouver could face a water-supply problem. However, in mid-November 2006, *The Globe and Mail* reported that residents were warned to boil any water to be used for consumption, after mudslides had stirred up water.

Confidence in the purity of the water supply had led officials to ignore a report that consultants submitted to the Greater Vancouver Water District Board in 1996, ten years earlier. It may be that the greater intensity of rainfall events that had been predicted as a result of climate change had not been factored into their decision. Since then, however, heavy rainfall has been clearly linked to increased risk of infection from pathogenic micro-organisms such as *Giardia* and *Cryptosporidium*, not to mention the *E. coli* outbreak experienced by Walkerton, Ontario, in 2002.

A study in the US linked heavy rainfall events to health impacts like these. The study covered the 48 mainland states over a fifty-year period, including both surface water and groundwater sources. Many regions and cities that had felt that their water was "safe" have more recently been re-evaluating the risk.

Hydroelectric Power

Eleven percent of Canada's energy mix consists of hydroelectric power; it is particularly important in the provinces of Quebec, British Columbia, Ontario, Newfoundland, and Manitoba. Large-scale hydroelectric power (say, more than a 20-megawatt capacity) may have significant environmental impacts, depending on the site. If forests are flooded to develop the reservoir, it is likely that methane will eventually be given off as the wood decomposes. Large dams may also displace local communities and disturb traditional hunting and trapping activities. However, compared with fossil fuels, hydroelectricity is environmentally benign. If hydroelectric capacity declined as a result of climate change, then, unless demand could be reduced, the use of fossil fuels would rise.

Three major types of adverse impacts can be predicted under climate change. First, extreme weather events, such as the 1998 ice storm, bring down hydro towers and interrupt supply; sometimes this takes place over a period of only a few days, sometimes for much longer. Second, lower lake and river levels reduce capacity; again there is a range—sometimes this takes place over a single season, sometimes over a medium-term cycle of about five years. And third, over the longer term, a warmer world implies greater losses in transmission.

There is also the distinct possibility that lower lake levels may be a permanent response to the new hydrological balance (we saw above that a new equilibrium could be reached between increased precipitation and increased evaporation). Although there is some uncertainty, we ignore the predictions at our own risk: in the words of Monirul Mirza, "most analyses for the Great Lakes suggest a decline in future lake levels and outflows" (Mirza 2004).

The 1998 ice storm exposes these new vulnerabilities. This storm, regardless of whether we see it as a climate-change event, illustrates the susceptibility of power lines to extreme and unexpected weather. In early January 1998 an unusual southern shift of the jet stream towards the Gulf of Mexico picked up a warm, moist air mass before moving onwards to the Ottawa Valley and Montreal, where it ran into a cold Arctic air mass. This air mass effectively slid underneath the warmer southern air, producing the phenomenon of freezing rain. As the warmer air mass rises it produces rain which then falls through the lower-lying cold Arctic mass; it turns to icy rain which then freezes on contact when it reaches the ground, buildings, and infrastructure. Ice can easily build up on power lines.

Of course, freezing rain is a regular phenomenon in Canada. But what made this particular event unusual was the fact that the weather system did not continue the usual westward progression into the Atlantic. Instead it was *blocked* by an extensive ridge of high pressure and became stationary for a period of six days. The storm event continued over an area running from Ottawa to Montreal, through eastern Ontario and parts of the Maritimes and New England.

Many of the hydro towers supporting the power lines collapsed under the weight of the ice. Four million people were without electricity for at least 36 hours, and some for much longer. Trees collapsed under the load of ice, and there was considerable traffic disruption, especially in Montreal. At the worst point, traffic could not leave the island of Montreal and the last functioning water filtration plant was down to *four hours* of supply. If a serious fire had broken out it would have been very difficult to control.

The number of deaths attributed to the ice storm came to twenty-two in Quebec and four in Ontario.

Forest Management

Just as the ice stores large quantities of water for Canada, so are there important benefits from frost. It protects people, animals, and trees from pests that cannot tolerate extended periods of cold. The pine beetle is one such pest.

This beetle can infest any type of pine tree, but has an established preference for the lodgepole pine, a species found in Alberta, British Columbia, and the northwestern United States. Lodgepole pine is known for its trunk, usually "straight with little taper": as such it is valuable for construction, railway ties, mine timbers, poles for power lines, as well as for supports for the lodges and teepees of native people—hence the name "lodgepole" (Farrar 1995).

Unfortunately, the lodgepole pine is vulnerable to the pine beetle. The spread of the beetle was controlled until the mid-1990s by a combination of cold winters and forest fires. In fact, the beetle is so voracious that it kills the trees in a region, which are then especially vulnerable to fires—an effective form of "negative feedback" that until recently kept the beetle in check. But as a result of warmer winters, the beetle has spread. Over half of the lodgepole pine in the centre of BC have been lost. Here, there is little debate: "climate change is unequivocally affecting the outbreak" (The Economist, 5 July 2008).

The great fear is that the pine beetle will cross the 100-kilometre gap separating it from the boreal forest and large swaths of jack pine that run from northern Alberta and the Northwest Territories right across Canada, through much of Ontario, Quebec, and the Maritimes (Farrar 1995). As the winds can carry the newly hatched beetles 300 kilometres a year, this dramatic possibility may be just around the corner.

Another challenge is the growing frequency of wildfires. A decrease in moisture from spring snowmelt and disappearing glaciers, and an increase in evapotranspiration, together with overall higher temperatures, all greatly elevate the risk of fire. Again, it is too early to state categorically that this is a trend driven by climate change. But given the seriousness of the situation, it would clearly be imprudent to wait another decade before taking action. Even the well-watered coast of British Columbia has suffered serious forest fires. In the interior, and on the Prairies, the situation is much worse.

News reports like the following are now commonplace:

> As wildfires carried thick black smoke too close for comfort to several communities in northern Saskatchewan yesterday, more residents were forced from their homes by the burning forests, bringing the evacuation roll to more than 2,000.

> So far this year, 460 wildfires have burned in Saskatchewan, up from 205 in 2007 and higher than the 346 fires calculated as the province's 10-year average for this time of year. (*The Globe and Mail*, 10 July 2008)

Containment of the fires requires the deployment of helicopters, tanker planes, and firefighters from across the country. These are not just expensive, but of course they also all emit more GHGs.

Transportation

Transportation will also be significantly affected by climate change, both from direct impacts and from the burden of GHG reduction.

One of the most serious impacts is a more active freeze-thaw cycle compared with what we have seen in the past, when many regions of Canada remained below freezing for several months without interruption. This cycle causes enormous damage to road surfaces, which are expensive to repair. The alternation between rainy and snowy conditions in the winter will likely require the use of more salt, while freight trucks will have to cope with higher winds and sudden downpours. Data from Ontario in fact show that this trend is already established.

And there is also the mitigation side. The transportation sector, responsible for approximately 30 percent of GHGs in Canada, will have to respond to pressure to reduce emissions. Fuel efficiency standards have been raised in most jurisdictions in recent years, in line with standards in other countries. Fuel switching is also part of the response, bringing hybrid vehicles, electric vehicles, and biofuels into the mix. Taxi fleets can reduce fuel use by improving vehicle maintenance; some are even aiming to earn carbon offset credits.

As with the electricity sector the most direct way to reduce emissions is to reduce demand. A good example of this is Toronto's solid waste disposal. The city had been using a landfill site in Michigan. This 800-km round trip was a huge waste of transportation capacity, and obviously generated significant emissions. By 2003, the garbage convoy consisted of 142 daily truck trips! Since then Toronto has bought a landfill facility in Ontario, but this is still 200 kilometres away. The new landfill, combined with an aggressive solid waste recycling program, brought the daily trips down to 74 by 2007. This amounts to more than 440,000 tonnes of waste every day. The city is aiming for a 70 percent diversion target by 2010, when the Michigan disposal contract expires.

Dramatic reductions in GHG emissions from transportation could be achieved by shifts in the "modal split"—the allocation of traffic by mode of transport. Big reductions in emissions from freight could be found by switching traffic from road to rail. In Europe it is common to move solid waste by special railway cars. The most obvious targets for reductions are urban commuters: 70 percent drive to work in Toronto. Typically, 80 percent of those vehicles carry only one person. Denser residential areas, for example, make public transport a better option.

The Saguenay Flood, June 1996

We have seen how the link between climate change and some extreme weather events has not been fully established. But other links are clear. Each of the following three examples fits closely with the expected changes in the hydrological cycle in a warmer world.

The first example is the Saguenay flood of 1996. The most readily identifiable cause of the flood was the rapid downpour of 19 and 20 July, which delivered 155 millimetres of rainfall in 50 hours. This was the equivalent of the usual rainfall for the entire month of July, or the same as the volume of water that flows over Niagara Falls in four weeks. Outcomes from this event included ten deaths, 16,000 people evacuated, 1,718 houses and 900 cottages damaged or destroyed, and $800 million in damage. Losses also included impacts on the pulp and paper industry and the aluminium industry, not to mention the effect on tourism at the height of the vacation season.

Like all major disasters, the causes of the Saguenay flood are multiple and complex. First, there had been considerable rainfall over the two weeks prior to the deluge, so the soils were already soaked and the reservoirs full. These "antecedent conditions" can usually be found before most major flood events (or even minor floods with major consequences like the Walkerton *E. coli* tragedy). Water levels in the dams were high. This was mainly because of local recreational demands, following three previous dry summers. The properties in the affected region included homes in small towns and many vacation cottages. Thirty years ago the cottages had been quite basic, without running water or electricity. Over time they had been gradually improved, and so was the infrastructure that served them. The value at risk to a flood increased over the years; this was especially true given that many of the cottages and the associated infrastructure had unwisely been constructed within the 20-year flood plain (i.e. land expected to flood once every 20 years according to historical data).

Responsibility for managing water levels in the 2,000 dams and dikes was fragmented between 25 different public and private organizations, with little coordination. As always, there is the conflict between holding water both as a reserve supply and for recreational purposes, and releasing enough water to ensure that the dams could offer flood protection when needed. Furthermore, like much of Canada's infrastructure, it was old and not built to current standards. When the dams overflowed and failed, the floods spread rapidly. Aside from mass evacuation, there was no back-up plan.

This was Canada's biggest flood disaster since Hurricane Hazel struck Toronto in 1954.

The Red River Flood, April 1997

The Red River Valley is a remnant of the post-glacial landscape described in chapter one. Specifically it is a part of the area that was covered by Lake Agassiz, a massive post-glacial lake that gradually drained away at the end of the last ice age 10,000 years ago. The exposed lake bed is a large flat expanse of land, rich in soils and very attractive for farming. This giant, fertile flood plain is both well populated and extremely vulnerable to recurrent floods.

Another key topographical/hydrological fact adds to the region's vulnerability: the Red River flows north, from the United States to Manitoba. Snowmelt begins in the south, at a time when the north is still covered with snow and/or ice. Given this reality, floods will be a regular spring occurrence.

A major flood in 1950 convinced the then-premier of the province that a heavy investment in a floodway would divert oncoming floodwaters away from Winnipeg. The floodway project was duly completed in 1968 at a cost of $63 million. At the time, many felt that this was a waste of money. However, even before the great flood of 1997, this floodway project had already protected Winnipeg from no less than 18 floods. So the Red River Valley is a very important case study for determining the costs and benefits associated with adaptation to climate change. Like the Okanagan Basin, the Red River provides an excellent laboratory for preparing Canada for climate change.

Weather conditions in the spring of 1997 provided a perfect combination for the flood of the century, the worst recorded since 1826. The valley was hit by heavy snowfall from 3 to 6 April; cold weather then set in to compact the snow. An early snowmelt in the south inundated Grand Forks, North Dakota, the centre of which was then burned to the ground in a fire caused by broken electrical circuits and leaking gas mains. (Grand Forks is only 200 kilometres upstream from Winnipeg on the Red River.)

Feverish efforts were made to build up the protective dikes, including a 20-mile extension to the southwest of Winnipeg, with a backup dike closer to the city limits. On 23 April 20,000 people were ordered to evacuate the valley; over 8,000 members of the armed forces arrived to support local emergency services—the largest Canadian troop deployment since the Korean War. Managing the water was difficult, and the flood eventually covered 202,500 hectares, about 5 percent of Manitoba's farmland.

Fortunately the floodway strategy and the emergency earthworks held up. Had they not done so, Winnipeg would have been severely damaged by the flood waters. It was estimated that without the floodway, 80 percent of the city would have been underwater and 500,000 people would have been evacuated. Disaster was averted; but it was close. If there had been just a little more rain or stronger southerly winds, then the embankments might not have held. In this case—even before the threat of climate change—planners and politicians made the right call.

Lessons Learned So Far/August 2005 Downpour/Hurricane Juan

It is only very recently that public opinion has come to terms with the reality of climate change in Canada. However, there remains little consensus on how to respond, either individually or collectively.

One difficulty in establishing a consensus is the uncertainty about the exact effect on the various Canadian climates. Many will remember the downpour in North York, in the north of Toronto on 19 August 2005. This downpour was part of a storm system that swept across southern Ontario that day, which included two tornadoes. In one hour, 103 millimetres of rain fell at the intersection of Yonge Street and Steeles Avenue—nearly twice the amount delivered by Hurricane Hazel in 1954. For fifty years Hurricane Hazel has been used by planners and engineers as the "design storm" for which infrastructure should be prepared; Hazel was referred to as "the 100-year storm." However, nature is clearly capable of much more.

We have to be prepared for surprises. Although the debate about how climate change might affect the frequency and strength of hurricanes continues, there is little doubt that the area traditionally covered by hurricanes is changing as the oceans warm up. For example, a North Atlantic hurricane maintained hurricane-force winds almost until it made landfall near Seville, Spain on 9 October 2005. Hurricanes had never been recorded in the South Atlantic until Catarina hit Brazil in 2004. Similarly, it is unusual for hurricanes to make landfall in the Maritimes given that the cooler Atlantic waters usually slow them down to tropical storm status before they reach the coast.

But Hurricane Juan was still a Category 2 hurricane when it passed between Prospect and Peggy's Cove, Nova Scotia, at midnight on 29 September 2003. The eastern "eyewall"—where the strongest winds occur—passed directly over Halifax Harbour. A hurricane like this had not occurred since 1893. Juan was the most damaging storm in the Maritimes in modern times, measured in terms of the number of trees blown down, power outages, and buildings damaged. Eight people died in the storm. The reason for this unusual event is attributed to the elevation in average temperature of the coastal waters, which was 18°C rather than the usual 15°C. This meant that the hurricane did not weaken as quickly as it normally would in northern waters. Juan in fact accelerated as it made landfall and maintained hurricane strength in its three-hour passage over Nova Scotia. It finally weakened to a "marginal hurricane" as it crossed the Northumberland Strait and hit Prince Edward Island. All this took place over a period of just four hours.

The warming of the surface of the oceans is one of the most predictable effects of climate change, so we should be prepared for more hurricanes outside of "expected" ranges. Beyond Canada, it is worth noting that Washington, DC, New York City, and Los Angeles all lie *just* outside the current extent of hurricane tracks.

The Canadian Hurricane Centre issued this statement in 2003:

> If the seemingly small departure of water temperature from 15°C to 18°C makes a significant impact on storm strength like we believe, then we should be very concerned about long-term trends in ocean temperatures (Fogarty 2003).

It is too soon to make precise regional forecasts regarding the specific impact of climate change on Canada. However, it is not too soon to accept that, on balance, the changes will be negative, and that some events may be locally devastating. As we have seen, traditional northern lifestyles are already headed towards irreversible change.

Given this reality, how has Canadian society responded so far? Let us return to the two broad questions raised at the start of this chapter. How can we adapt to the changes to which we have already committed ourselves? And how quickly can we move to reduce GHG emissions?

Chapter 7

The Political Arena

It is less than realistic for 13 jurisdictions to run off in 13 separate directions inside one country. We need strong federal leadership. We'll do everything we can on our own, but it is less than ideal.

Dalton McGuinty, Premier of Ontario, 19 July 2008

The bottom line is if the US moves, does Canada stand still? I don't think they can.

Gary Doer, Premier of Manitoba, 19 July 2008

The Global Context

The global quandary

From one perspective the climate change problem is simple: we know that we are at risk, and that we must immediately work to minimize this risk. But as soon as we being to think about how to operationalize this, problems arise. We can see these problems as technical, economic, and moral, or political—recalling Adam Smith's "moral philosophy."

On the technical side there is no magic bullet. There are many emergent technologies, but these will take time. Also, the concept of "technology transfer" has changed since it first emerged in the context of international development: it turns out that technology is not necessarily always transferred from richer countries to poorer ones. At the moment, the third-largest wind turbine company in the world is Chinese. Still, the poorer countries in the developing world will need support to develop low carbon economies, in the same way that subsidies are available in richer countries to encourage new technology.

A major issue here is competitiveness. Will the countries that reduce their GHG emissions be disadvantaged? This issue was a key influence on the design of the European Union Emissions Trading Scheme. We saw how the major burden of GHG reduction was imposed on electricity producers, given that their sector was less subject to international competition than, say, producers of aluminium or steel. Now that climate change is back on the US federal political agenda, the impact of GHG reduction targets on competition has become a point of serious debate.

It is in fact another aspect of the "upstream-downstream" debate, or, more informally, "whose carbon is it, anyway?" In the longer run it does not matter where the initial "carbon burden" is placed, because the effects will diffuse up and down the energy chain, the supply chain, and all the other chains that link the global economy. However, in the short run, it matters very much. Wherever the burden is placed first—whatever country, or sector of the economy, or supplier, or consumer—will have to adjust to living in a "low-carbon world" before everyone else. Hence the protests.

The US has long been absent from the Kyoto process. But even so, the Kyoto process, the debate in Congress on the Waxman-Markey climate proposal (the American Clean Energy and Security Act) focussed on this issue: how could the free-riders (those without GHG reduction targets) be forced to pay for their GHG emissions, so that America would not be at a competitive disadvantage? The same debate is ongoing in the European Union (EU). Here France championed the idea that free-riders should pay a special duty on their exports to the EU. This is a sensitive political issue, but it is no excuse for inaction. The potential losses from being disadvantaged economically by moving first on GHG reductions are *trivial* compared with the potential costs of climate change impacts. Ask the inhabitants of New Orleans who are still unable to return home several years after Hurricane Katrina struck.

These moral (or historical/political) issues are not yet so prominent in the international public debate, although they have always been present in the background. The rich nations (in terms of wealth per capita) became rich, to a large extent, by unleashing the power of fossil fuels. They began to emit excess GHGs that accumulated in the atmosphere at a time when the poorer nations were still reliant on biomass, wind, water, and somatic energy (the bodily energy of people and animals). There are two big questions:

- How can the rich countries expect the poorer countries to cut their GHG emissions while the mass of their people are still poor?
- What point is there for richer countries to cut their own emissions while major economies (like India, China, Brazil, Mexico, and Indonesia) have entered a period of rapid economic growth based on fossil fuels?

There is an elegant—although optimistic—answer to this dilemma called "contraction and convergence." In this scenario, worldwide emission targets fall to the same per capita amount. The shorter-term pragmatic response is that the richer countries have a moral and practical duty to move first, to show how the world's transition to a low carbon economy can begin. It has to start somewhere.

The Kyoto process

The 1992 World Summit on Sustainable Development in Rio de Janeiro was the start of the global response to climate change. The summit produced the United Nations Framework Convention on Climate Change (UNFCCC) from which it was expected that a protocol on climate change would eventually emerge. The framework was created for the annual meetings of the "Conference of the Parties," which in turn led to the signing of the Kyoto Protocol in 1997. Under the Protocol, the signatories from the richer, industrialized countries agreed to cap their GHG emissions at some percentage below the latest available baseline for emissions, 1990, by the end of the First Compliance Period (2008–2012). For example, the US pledged to go 7 percent below the baseline by this target date; Canada pledged 6 percent below, and so on.

The Protocol would come into force when it had been adopted by at least 55 of the capped signatories, accounting for at least 55 percent of their total GHG emissions. The US withdrew from the Protocol in 2001; Australia withdrew in 2002, the same year that Canada ratified the Protocol. In December 2004 Russia, a big GHG emitter, finally ratified the Kyoto Protocol, making up for the absence of the US. The Kyoto Protocol came into force in February 2005.

The principal reason given by the US and Australia for their withdrawal was the fact that poorer countries—particularly major emitters like China and India—had not accepted caps. The richer countries saw this as an economic competitiveness issue; for the poorer countries it was primarily a political or moral issue. Their economies needed time to catch up with the richer countries. This impasse has yet to be resolved.

In the meantime the Kyoto Protocol continues to evolve as signatories have struggled to activate the major elements crafted by the annual meetings of the Conference of the Parties. Some of these details will be explained in the rest of this chapter and subsequent chapters.

It is important to recognize that "Kyoto" is not the only game in town. The US and Australia formed the Asia-Pacific "Group of Six" with China, India, South Korea, and Japan to maintain a dialogue on response to climate change. Canada's federal government announced an interest in joining this group, having declared that Canada's Kyoto commitment was unattainable. Climate change has been discussed by the G8—the US, Japan, Germany, the UK, France, Italy, Canada, and Russia—as a leading issue for their annual meeting, although little action has emerged so far. The World Bank has been an early advocate of action in developing countries, in

2000 establishing the Prototype Carbon Fund. This was the first "carbon fund" for investment in climate-friendly projects. Many such funds are now in existence.

The Clean Development Mechanism and Joint Implementation

From the earliest debates following the Earth Summit in Rio it was clear that there had to be some way to involve poorer countries in the process. Surely it was possible to draw these economies away from the fossil fuel pathway that had made rich countries rich. How could elements of a low-carbon economy be introduced in poorer countries, when those countries lacked caps on GHGs and/or investment capital for clean technology? A means was needed to encourage richer countries to fund these projects, funding that did not rely solely on altruism. There was also an economic argument: there should be projects in poorer countries where it would be *cheaper* to reduce GHGs than it would in richer countries (which presumably were more efficient users of fossil fuels).

All of these arguments and aspirations lay behind the creation of the Clean Development Mechanism and Joint Implementation as important elements of the Kyoto architecture.

The Clean Development Mechanism (CDM) links investors in GHG-capped countries to mitigation projects in uncapped countries. For example, a Canadian investor or a Canadian government invests in a Chilean project to capture methane from pig manure. One objective of the project is to *reduce* existing GHG emissions (in this case methane) in Chile. Another type of project is based on *avoiding* future GHG emissions. For example, another Canadian investor might support a small hydroelectricity project in Costa Rica, arguing that if the project had not been developed, then additional power would have been provided from fossil fuels.

In order to have clarity on how much GHGs have been reduced or avoided, every CDM project must follow an approved protocol, or method of analysis, which is then verified by an independent third party. This third party consists of a professional organization (also approved by the executive board of the CDM). Clearly there could be considerable disagreement over the quantity of GHGs avoided, as compared with a hypothetical alternative. But once all these hurdles have been cleared, then a particular CDM project can be accepted as being worth a given number of credits per year over the lifetime of the project. Each project is continually inspected by the verifying agent to ensure that it is functioning properly and that the ensuing CDM credits are genuine. These credits have value and may be transferred from the original investor, or "originator." Hence there is now an active global market in CDM credits.

An observer might be tempted to think that such a complex scheme would be doomed to failure. Indeed it did get off to a slow start; however this was largely a result of the executive board being grossly underfunded. But the CDM is now up and running.

Joint Implementation is a similar system set up for investment by a capped country in another capped country, where emissions reductions might be achieved more cheaply than domestically. For example, a Canadian investor or government might invest in a combined heat-and-power plant in the Ukraine or a wind farm in New Zealand, obtaining tradable credits just like the CDM. Although the Joint Implementation mechanism was proposed before the CDM, it was even slower to start up than the CDM.

It is too soon to claim that the Clean Development Mechanism and Joint Implementation systems will achieve all that is hoped. However, clearly this ingenious scheme to bring uncapped countries into the challenge is cause for optimism.

The European Union Emissions Trading Scheme

It is often said that Canada's federal structure makes decision-making extremely complex compared to other countries. It should be very instructive, then, to see how the European Union developed a climate change policy in a much more complex political and cultural framework with 25 official languages, and varying in size from Germany with 83 million people to Malta with 300,000.

At the Kyoto conference the European Union (EU) adopted a collective target of 8 percent below the 1990 baseline on the understanding that individual targets would vary from country to country. This would allow the poorer countries to increase their emissions, and the richer countries to take deeper cuts to compensate. The EU then developed an Emissions Trading Scheme (EU ETS) to begin implementing the objectives and instruments of the Kyoto Protocol in a "Phase One" from January 2005 to December 2007, as a practice run before Kyoto's First Compliance Period from 2008 to 2012. Of the six GHGs, only CO_2 was included in Phase One.

The target group for emission caps were all installations with a generating capacity of 20 megawatts or more among power producers, oil refiners, ferrous metal producers, and manufacturers of pulp and paper, cement, glass, and ceramics—more than 11,000 in total. Each EU member country set up a National Allocation Plan, with overall targets set by the EU Environmental Commission, leaving each member country responsible for allocating the national cap among its installations. This cap-and-trade system (described in chapter five) allowed each installation to either make the required cuts in emissions or buy a credit from another participant who had bettered its target, a strategy known as "make or buy." Since the market opened in January 2005, the volume of trading in "carbon credits" has grown steadily. But have emissions been reduced? The answer, not surprisingly, is "not yet." The first three years of the European ETS is too short a time horizon to work into the investment cycle for lower-carbon power sources, such as a switch from coal to gas (although we have seen that this shift was happening anyway in Britain through the 1990s).

Another lesson from Phase One is that the caps were set far too high. At the end of the first truing-up, for the year 2005, emissions allowances amounted to 1,829 million tonnes of CO_2. But the actual emissions were only 1,785 million tonnes, creating a surplus of credits of 44 million tonnes. As news of this surplus leaked out before the official statistics were published, the price for Phase One carbon credits collapsed.

It would also help the development of the carbon market if some of the credits could be auctioned rather than distributed for free, as they all were. Obviously, there was political value to be gained from starting the scheme with a free distribution. A further lesson was that it is worth taking extra time to independently verify the baseline emissions for each installation, instead of waiting for this to be done at the conclusion of the first year of trading.

None of these deficiencies have proved to be terminal. Lessons were learned and the emission caps for Phase Two are much tighter. The EU has also added a renewable energy target for each member state as an extra incentive. Of the additional "mechanisms" for trading, only the CDM was used in Phase One; Joint Implementation is included for Phase Two and discussions are underway to bring aviation emissions into the scheme, as well as the other five greenhouse gases. One important issue to be settled was the extent to which CDM credits could be counted towards a national reduction target. There was some jockeying between the EU Environmental Commission and the member states on the number. Ireland suggested up to 50 percent. This was clearly too high, and rejected by the EU, although it did leave the member states to make an offer. Most states finally settled for a number between 6 and 10 percent.

Nobody is saying that because of this scheme, the battle against climate change is over. However it is encouraging to see how a large-scale GHG management plan might be implemented.

Canada and the international scene

By and large, Canadians believe that they take environmental issues seriously. Certainly Canada played a leading role at the Stockholm Conference on Human Environment (1972) and also The Brundtland Commission on Our Common Future (1983–88). Environment Canada hosted "The Changing Atmosphere" conference in Toronto (1988). Maurice Strong was a major force behind the Earth Summit (1992) as well as the global initiatives that the summit spawned, including the

Kyoto Protocol. Furthermore, as a northern country stretching into the Arctic, Canada has many physical features in common with the Scandinavian countries—so perhaps some of their enthusiasm for environmental standards is linked to Canada by association.

Following the Earth Summit, the Conference of the Parties process began to shape what would become the Kyoto Protocol in 1997. Certainly, at one of the early Conference of the Parties meetings (Berlin in 1994), Canada played an active role. However, although the supportive rhetoric continued after Berlin, the momentum of action seemed to fade until it disappeared altogether. Despite broad cross-party support to ratify Kyoto, it took five years for Parliament to pass the ratification bill. Even as the science of climate change became increasingly specific about the risks of inaction, federal and provincial governments began to worry about emission caps being "bad for business."

The origin of this discomfort is somewhat unclear. It certainly did not come from the business sector, where the GHG issue was well understood, especially among the major emitters (many of whom had already accepted the inevitability of caps on GHG emissions). Many business people, the professions, consultants, and some labour unions, saw that inaction would be far more expensive than exploring the new opportunities of a carbon-constrained world. (We will come back to this in chapter eight.)

The election of George W. Bush was a major setback for action on climate change in Canada as well as the US. Despite the clear scientific consensus, there was a mindset in the United States (and to some extent in Canada) that could not accept that the leading industrial economies should take the initiative in GHG mitigation, without caps being placed on the world's poorer countries. In the US, this feeling was directed particularly against China, which was increasingly feared as an important competitor for global markets, especially in manufacturing.

Prime Minister Jean Chrétien announced at the 2002 Earth Summit in Johannesburg that his government would introduce a bill to ratify Kyoto the following spring. And so it was done. The new Liberal government, under Mr. Martin, lasted only a few months. Under the Liberals nothing of any significance was done to implement the Kyoto Protocol in Canada. Stephen Harper's minority Conservative government soon declared that the Kyoto target was unattainable.

In 2006 the federal government introduced Canada's Clean Air Act, which was based on an "integrated approach": in this, GHG emissions and air pollutants were linked together for action. The details would be specified over a three-year period.

Ironically, one Kyoto requirement that successive federal governments did undertake was the publication of a National Report on Climate Change every four years (this had been promised under the UNFCCC at the Rio Earth Summit). *Canada's Fourth National Report on Climate Change: Actions to Meet Commitments Under the United Nations Framework Convention on Climate Change* was submitted in 2006. The latest version makes disheartening reading for anyone looking for significant action on climate change. Surprisingly for a government document, on page 1 it presents a very blunt statement on the true state of affairs:

> Even though the federal government had announced billions of dollars in funding since 1992 towards meeting commitments to address GHG emissions, as of 2004 Canada's GHG emissions were 26.6 percent above 1990 levels. The Commissioner urged Canada's new government to come up with a credible, realistic and clear plan that should address the long-neglected need to help Canadians cope with the consequences of climate change and to commit to specific actions with timeframes for completing them.

The 300-page-long report goes on to describe various measures that will be taken and funds that will be committed to activities such as increased reliance of public transport, energy efficiency, targets for renewable energy, the closing of Ontario's coal-fired power plants, biofuel requirements for gasoline, and so on. But there are no caps on emissions and only a promise that over the next three years agreements will be made with companies in the major emitting sectors—buildings, transportation, industry, agriculture, and forestry.

The second page of the report admits that

> from 2010 to 2020 emissions from the upstream oil and gas sector are expected to decline somewhat, as conventional oil production declines, while emissions from refining and synthetic crude oil production (i.e. oil sands) will continue to *increase*. Emissions from all other sectors (with the exception of electricity) are expected to *increase* further, notably in the transportation, industrial, residential, and commercial sectors. (Emphasis added.)

Amazingly, this stark admission of inaction is immediately followed by three paragraphs on the "Physical and Socio-Economic Impacts," which concluded that

> these could have serious ramifications for agriculture, tourism, municipal water supplies, water transportation and wildlife habitat. The forestry and fisheries sectors could also be threatened by possible changes in climate. (Report available from Environment Canada at www.ec.gc.ca.)

Any Canadian who attends an international climate-change conference should expect to be asked what Canada is doing on the issue. The answer unfortunately is that, even after all the promises and publications, Canada's GHG emissions are still going up, and there is no concrete action yet in place.

The United States

It is clear that US action or inaction has a major effect on Canada's climate change policy. And while it is tempting to blame the eight years of George W. Bush's presidency for lack of progress, the situation is actually much more complicated. We can start by looking back to the presidency of George Bush Sr.

At the Rio Earth Summit it was said that George Bush Sr's instructions to the American delegates were as follows: they could agree to target dates or to target emission cuts, but not to both. US opinion on climate change was more polarized than in Canada. The US was home to the business lobby called the Global Climate Coalition, a network of large corporations from the oil and gas, chemical, and automotive industries. The coalition denied the existence of climate change, and vigorously opposed any action to reduce emissions. They even made a movie claiming that a CO_2-enriched atmosphere would be so good for plant life that the Sahara Desert would soon be green with vegetation!

At the other end of the spectrum were groups like the Boston-based Union of Concerned Scientists, many of whom were key members of the Intergovernmental Panel on Climate Change (IPCC). They began lobbying the federal government for action in the early 1990s. One of George W. Bush's first actions on climate change was to carry out an American appraisal of the risk, calling on members of the US National Academy of Sciences. Apparently his advisors did not know that American scientists were by far the largest group within the IPCC itself. Furthermore, Boulder, Colorado is the home of the National Center for Atmospheric Research, which produces some of the leading climate change science. Predictably, the National Academy fully endorsed the work of the IPCC.

Interestingly, the US was the first country to use market forces to reduce pollution, drawing on amendments to the Clean Air Act to reduce sulphur dioxide and nitrogen oxides (the cause of acid rain) from 1994. Following the program's overwhelming success, US negotiators at the early Conference of the Parties meetings proposed it as a model for what would become the Kyoto Protocol. Other representatives, especially the Europeans, were skeptical and continued to propose a traditional command-and-control framework to reduce GHG emissions. Eventually the American view prevailed. Then President Bush withdrew the United States from the process.

By this time, a shift was already underway in American public opinion. The shift was most clearly apparent when John Browne, then CEO of British Petroleum, made a speech at Stanford University in 1997. Climate change was a significant threat to the planet, said Browne, and GHG emissions must be reduced immediately. He withdrew BP from the Global Climate Coalition and set up an internal carbon credit trading scheme within the company. BP also embarked on major investments in solar power. The coalition began to break up, with Royal Dutch Shell, Dow Chemical, and others soon following BP's exit.

Insurance companies in the US had rejected the European insurers' support for action on climate change, led by the world's largest reinsurers, Munich Re and Swiss Re. But even then a major exception was AIG, America's largest insurance company, which was an early proponent on action.

Leadership was also provided by the Chicago Climate Exchange (CCX), set up within the Chicago Board of Trade, to develop a voluntary market for trading carbon credits among major emitters and providers of carbon offsets in 2003—two years before the European ETS. Founding members included Monsanto, Dow Chemical, American Electric Power, and Ford Motor Company. Although this is a voluntary initiative, members are committed to legally binding targets for reducing GHGs.

Throughout America from the mid-1990s onwards there was growing evidence that many people took seriously the threat posed by climate change. We have already seen grassroots initiatives at the municipal level. For non-governmental organizations (NGOs), climate change became a major issue. Interest at the state level grew, beginning with California. A series of networks were formed covering more than half the country, particularly the Western Climate Initiative and the Regional Greenhouse Gas Initiative in the northeast. In 2008 climate change became conflated with the rising cost of oil imports and the national energy security issue. Even George W. Bush took note, announcing subsidies to develop biofuels, specifically corn-based ethanol. American oil and gas imports amount to $800 billion per year. The legendary Texas oilman, T. Boone Pickens, announced a $2 billion investment in wind power in the Texas panhandle. In the summer of 2005 a letter signed by over 40 senators from both parties urged Bush to take action on climate change.

It was an event later that summer that may have permanently ended any lingering doubts about the risks of climate change.

On 5 August 2005, Hurricane Katrina struck New Orleans and the Louisiana coast. Over one thousand people died; 300,000 people were evacuated. Insured losses were initially estimated at US $27 billion (later revised to $40 billion), and total economic losses at $100 billion. Of the evacuees, half had not returned three years later—their homes no longer existed, could not be insured, and so could not be rebuilt. The courts are still hearing cases on the issue of the storm surge (the seawater rushing onshore in the wake of the hurricane), which accompanied the hurricane's landfall. Damage from storm surge was estimated to be responsible for more than 75 percent of the total damage and is excluded from a standard home-owner's insurance policy. In compensation, insurers typically offered only about 25 percent, if not less, of the replacement value. The destroyed homes and businesses cannot be rebuilt in their original location given that they were below sea level. Today, these houses would not be given planning permission.

There is a final irony in this great tragedy. The hurricane struck during the week that the National Association of Insurance Commissioners (NAIC) was expecting to hold their annual meeting—in the city of New Orleans itself. Members of the NAIC were among the leading skeptics on climate change. After this event, they publicly urged President Bush to take action on climate change.

In 2008 California passed legislation excluding the importation of high-carbon fossil fuels, such as Canada's oil sands. There is similar legislation before Congress prohibiting the use of "dirty" fossil fuels (the oil sands again) by federal agencies such as the post office and the armed services. Now that concern has spread widely through the US we are back to the old question: Whose carbon is it, anyway? Is it the producer's carbon, or the consumer's?

Other Kyoto signatories: New Zealand, Japan, and Australia

Canada has not been alone in its struggle to develop a coherent approach to climate change. Apart from European Union countries, until recently New Zealand and Japan were the only other countries, aside from Canada, that had accepted caps under the Kyoto Protocol.

Japan accepted a 6 percent reduction below the 1990 baseline (like Canada) while New Zealand has agreed to return to the 1990 level of GHG emissions. Although Japan and New Zealand have not done as poorly as Canada's 27 percent increase in emissions, neither have they implemented a plan to begin the reduction process. Like Canada, they have produced plenty of plans and promises of more to come.

Perhaps there are lessons to be learned for Canada from Japan and New Zealand's inability to move forward on GHG reductions, or even to implement a schedule to start. For example the three countries share a key feature: isolation. When the US and Australia withdrew from Kyoto (in 2001 and 2002, respectively), Canada and New Zealand lost their best opportunity to set up a regional/international carbon trading system akin to that in the European Union. For Japan the isolation stemmed from the fact that none of its nearest trading partners, such as South Korea and China, had accepted caps under Kyoto.

New Zealand passed the Climate Change Response Act in 2002, creating the legal framework for GHG inventories and registries that would enable trading of carbon credits on the international market. It has also established the necessary elements for a national carbon trading system. However, as a small economy, New Zealand has minimal opportunities for trading. Many positive actions have been taken at the local level under the Communities for Climate Protection program to reduce emissions and strengthen stormwater infrastructure in expectation of a more intense rainfall regime. However, the targeted return to the 1990 emissions level remains elusive.

Although New Zealand does not face the massive shift in hydrological resources projected for Canada's loss of snow and ice, the future is uncertain given the economy's dependence on agriculture and tourism. Not only are both sectors highly susceptible to the physical impacts of climate change, but they are also susceptible to declines in international demand. The potential problem lies in New Zealand's distance from most of the rest of the world to which it exports its produce and from which it welcomes its tourists. In a carbon-constrained world, traffic with New Zealand might be deemed to come with too big a carbon footprint.

Japan's *per capita* GHG emissions are less than half those of Canada. Like Europeans, Japanese people are used to limited space. The rate of car ownership is much lower than in Canada. However, the household trend is to consume more, combined with increasing car ownership. It is estimated that household GHG emissions have risen by 40 percent since 1990. Japan has been very active in investing in CDM projects (especially in the region), but so far it has not imposed caps on industry. Instead for the most part industry sets its own voluntary targets. Japan is now 6.5 percent over its 1990 baseline.

So far, it is Australia that has gone through the most dramatic policy swings of any of the Kyoto-capped countries. Under the Conservative government of former Prime Minister John Howard, Australia followed George W. Bush's example and withdrew from the Kyoto Protocol in 2002, despite widespread support by state governments (all of them governed by the Labour Party). Indeed one of the earliest mandatory carbon trading schemes in the world was set up on a trial basis in New South Wales in 2003—the New South Wales Greenhouse Gas Abatement Scheme. Subsequently the states established a National Emissions Trading Scheme.

Australia is extremely vulnerable to climate change; it is already exceedingly prone to drought, even in the better-watered parts of the country. The northern halves of Australia's Indian Ocean and Pacific Ocean coastlines also lie within a hurricane zone. Like Canada, Australia is poised to exploit a growing demand for its fossil fuels, in this case coal for the growing markets of Asia. As with Canada, Australia faces the question: "Whose carbon is it, anyway?" It is an issue that won't go away.

A national election in November 2007 replaced the Howard Conservatives with a Labour government headed by Kevin Rudd, who promptly signed a ratification of the Kyoto

Protocol. Furthermore he declared that Australia's Kyoto target was still attainable.

In some ways, of these three comparisons Australia's is the most relevant to that of Canada. It has a federal system; the states took initiatives independently—indeed quite opposite—of the federal government; and it is starting very late to meet a Kyoto commitment.

All three of Canada's Kyoto-capped non-EU comparators have failed to move forward. Each has constructive lessons for Canada. Especially from Australia we can learn that even if GHG mitigation is a bottom-up initiative from the states or provinces, the federal government can still catch up, provide an integrative framework, provide renewed leadership (following an election), and not give up on the Kyoto targets. Like New Zealand, we can take heart that once our large neighbour comes back to the fold, then we can also return to the Kyoto table. From Japan, we learn that relying on voluntary efforts from industry will not be enough.

The Canadian Domestic Process

Northern Aboriginal communities

We have seen how the impact of climate change on the north will be more severe than anywhere else in Canada, because this is where the greatest warming will take place. This is also the land of snow and ice, and is the most vulnerable to warming. Furthermore many of the inhabitants are already among Canada's most vulnerable people.

> Just over half of the approximately 100,000 northern residents are Aboriginal and belong to distinct cultural groups including Yukon First Nations (Yukon), Dene, Métis and Gwich'in (Northwest Territories), and Inuit (Nunavut, Nunavik, the new Inuit land claim area of Nunatsiavut within the region of Labrador and the Inuvialuit Settlement Region of the Northwest Territories). Many of the communities are characterized by an increasingly young and rapidly growing population. (Furgal and Seguin 2006, 3)

For the Aboriginal population, even now, life expectancy for both men and women is 12 years below the national average. Most communities lack access to basic health services and clean drinking water. Another factor that makes them particularly vulnerable to climate change is their dependence on food from the land, lakes, rivers, and seas. "More than 70 percent of northern Aboriginal adults harvest natural resources through hunting and fishing and of those, more than 90 percent do so for subsistence purposes" (Furgal and Seguin 2006, 3).

The changes to northern ecosystems pass directly through the wildlife food chain to the Aboriginal population, adding to the insecurities and contamination already implanted by airborne pollution from the south. The unpredictability of the weather adds to often risky hunting, with ice breaking up early, storms arising suddenly, and prey, such as seal and caribou, moving to new areas or suffering dieback from disease or lack of forage.

Adapting to these difficult conditions is worsened by the slow pace of treaty settlement between the government and Aboriginal people. Ironically, this situation will become only more difficult as the pace of oil and gas development picks up to take advantage of the greater accessibility to offshore resources provided by the reduction of land and sea ice.

Federal and provincial initiatives

The opening quote of this chapter provided a warning: "The decision now facing Canada is significant. It is also an agonizing one, because it raises all of the regional and continental aspirations and conflicts that have plagued the Canadian federation since 1867" (Macdonald et al. 2002, 1). As we have seen, these words were prophetic. In the government's own report to the UNFCC it admitted that, "The Commissioner urged Canada's New Government to come up with a credible, realistic, and clear plan that should address the long-neglected need to help Canadians cope with the consequences of climate change and to commit to specific actions with timeframes for completing them" (Government of Canada 2006b).

We have seen the history of GHG mitigation at the federal level—many plans under two Liberal prime ministers over a period of more than ten years that resulted in imperceptible action on the ground. The Kyoto Protocol was ratified, but this ratification has turned out to be meaningless. While reports have been filed and promises have been made, the federal government has failed to make any substantial progress.

Finally, we have been witnessing some action at the provincial level, with climate change legislation passed in Alberta, British Columbia, and Quebec, and promises from Ontario, Saskatchewan, and Manitoba. Ontario and Quebec have signed a memorandum of understanding to cooperate on climate change and energy issues, including setting up a GHG cap-and-trade system. Ontario, Quebec, British Columbia, and Manitoba have also joined the Western Climate Initiative, which was formed by a coalition of US states, led by California. (The most up-to-date version of the plan is available at www.westernclimateinitiative.org.) Another six American states, six Mexican states, and Saskatchewan have observer status at the Western Climate Initiative.

Together with the northeastern US states in the Regional Greenhouse Gas Initiative, we have the beginning of a bottom-up movement towards a continental GHG trading system.

Municipalities and non-governmental organizations

It is good that some of the senior levels of government are finally taking action. However, it should be noted that the earliest actions were taken at the grassroots levels by NGOs, municipalities, and some businesses (more on this in the next chapter). Organizations such as the World Wildlife Fund, The Sierra Club, The Pembina Institute, The David Suzuki Foundation, Pollution Probe, and professional institutes such as the Canadian Environmental Network and Canadian Institute for Environmental Law and Policy have all been active since the late 1980s. New NGOs, like Zerofootprint and The Carbon Reduction Fund have been formed to deal with specific aspects of climate change.

Municipalities have also been active on an individual basis since the 1990s, and since 1994 through the International Council for Local Environmental Initiatives. The first targets for municipal action are reducing the emissions generated by municipal activities, including those of suppliers and contractors. However it can be difficult for urban governments to affect the emissions of the population at large, given that the instruments enabling them to do so are often in the hands of senior governments. Many people at the urban government level are acutely aware of both the inputs into GHG emissions (such as daily traffic congestion) and the sometimes deadly outcomes (such as severe heat waves).

Conclusion

The political arena in which the drama of climate change is being played out is a very complex place. The twin challenges of mitigation and adaptation produce tensions at every political level. Internationally, the tension is well-known. How much initiative will the richer countries take before the poorer countries with large economies (India, China etc.) accept caps? Is the carbon burden going to fall first on producers or consumers, on importers or exporters?

On a continental level how will climate change affect relations between Canada, the US, and Mexico? Given the cross-border difficulties we have had over relatively simple issues (fisheries, softwood lumber, and BSE), how will climate change play out?

Domestically, we have already seen the unfortunate consequences of failed federal leadership. Instead of climate change becoming a cross-party initiative it has become a political football. And we cannot even agree on the shape of the ball. Interprovincial rivalries have resurfaced along the lines of the infamous National Energy Program of the Trudeau era, still reviled in the west. Saddest of all, we have seen how climate change will undermine northern lifestyles, in spite of the fact that northern Aboriginal people have the lowest carbon footprint of all Canadians. As the opening quotation to this chapter promised, climate change "raises all of the regional and continental aspirations and conflicts that have plagued the Canadian federation since 1867."

Chapter 8

Business and the Economy

As an investor, Caisse de dépôt et placement du Québec attaches a great deal of importance to greenhouse gas emissions and corporate carbon management strategies. That is why the Caisse is a signatory to the Carbon Disclosure Project. The Caisse believes that climate change is a real issue and that if nothing is done, its impact will be monumental in the long run.

Henri-Pierre Rousseau, President and CEO,
Caisse de dépôt et placement du Québec

The economic risk of climate change has moved it from the realm of scientific debate to the front lines of investment risk management. Deloitte is proud to be the lead sponsor of the Carbon Disclosure Project: Canada 200 Report *initiative, which helps governments, companies, and shareholders develop rational responses to climate change.*

Alan MacGibbon, Managing Partner and Chief Executive,
Deloitte & Touche LLP (Canada)

Shares in auto parts suppliers have tanked as consumers swore off gas guzzlers, but some companies focusing on green technologies have a better chance to emerge as leaders.

Steve Ladurantaye, *The Globe and Mail*, 29 July 2008

Carbon Finance in the Context of Environmental Finance

Our belated response to climate change can be explained partly by skepticism about the seriousness of the threat.

Even when that threat became better understood, the reluctance to act persisted because many decision-makers assumed that reducing GHG emissions would place a significant burden on the economy. This, it was felt, would be especially true if one country moved faster than its competitors. Perhaps only with the publication of the *Stern Review* in October 2006 did it become clear for many that the probable costs of inaction *far* exceeded the likely costs of implementing an effective GHG reduction strategy.

In the meantime many business leaders had already accepted the seriousness of climate change. They prepared to take action while chiding governments for doing so little, and this inaction perpetuated uncertainty about the structure and timing of climate-related regulation.

During discussion of a planned $3.3 billion coal gasification facility near Edmonton, for example, the chairman of the board of Sherritt International explained:

> The absence of clear and articulated rules and regulations and solutions for greenhouse gases has the effect of absolutely stopping investment in its tracks.
>
> There has been an absolute lack of leadership on both the provincial and federal level.... They are sticking their heads in the sand. (*The Globe and Mail*, 23 May 2008)

Opinion had begun to shift not only among business leaders. At the same time, economic instruments were being developed that were designed to enlist the power of the market to drive innovation. This was one of the reasons why some far-sighted business people began to look for "opportunities in the greenhouse." In the late 1990s the term "environmental finance" emerged to describe this new thinking. This term encompasses "all market-based instruments designed to deliver environmental quality and transfer environmental risk" (Labatt and White 2002). Carbon finance is a subfield of environmental finance that specifically addresses climate change; "carbon" is used here as the short form for all GHGs (Labatt and White 2007). The hope was that markets would be developed by regulations that capped emitters while giving them the opportunity to compete with other emitting companies to find the cheapest solution. Once competition was underway a market would develop with transparent price discovery.

This breakthrough in innovative thinking about links between markets and environmental quality started with the establishment of the US acid rain reduction program in 1990. At this time, the Clean Air Act was amended to place caps on the emissions of sulphur dioxide (SO_2) and nitrogen oxides (NO and NO_2) from power stations. This began with Phase One in the Ohio Valley and in 2000 went nationwide in Phase Two. This was the earliest large-scale application of the cap-and-trade approach to reducing harmful emissions. Soon a lively market developed in surplus credits earned for reducing emissions below the caps placed on each designated installation. More importantly, as the caps were tightened, SO_2, NO, and NO_2 emissions began to fall. It was because of the evident success of this program that US negotiators in the early Kyoto meetings advocated a similar approach for reducing GHG emissions, as we saw in chapter seven.

The cap-and-trade approach has become the centrepiece for reducing GHG emissions under the Kyoto Protocol and the EU Emissions Trading Scheme. But it is also important to understand that there are significant differences between GHGs and acid rain. The key chemical difference is that the gases causing acid rain have a much shorter "residence time" in the atmosphere, usually only a matter of days; this is very different from the decades spent aloft by GHGs. In other words, acid rain gases are deposited fairly closely to their point of emission. Although some do travel internationally, sometimes crossing major oceans, it is possible to know which emissions came from which source. They can be "fingerprinted" from the known composition of fuels used by each plant. As well, real-time monitors are placed in every chimney stack. So to a large extent the US could tackle *its own* emissions, knowing that it would reap the benefits.

Clearly acid rain is a different problem altogether from climate change: GHGs spend a long time in the atmosphere and every country's emissions become mixed together. The only solution is an international one.

Another major difference between acid rain and climate change is that the technological choices for SO_2, NO, and NO_2 reduction were well known, available, and affordable within the investment cycle. For example, coal with lower sulphur content was available within the US albeit carrying a higher transportation cost than locally available coal. Or new plants could be built as combined-heat-and-power, or power generation could be switched to cleaner natural gas. Furthermore "scrubber technology" was already available for capturing unwanted emissions in the smokestack. This is very different from new technologies like carbon capture and storage, still unproven on a commercial scale and with costs of implementation unknown.

This is not to say that the acid rain market model cannot be applied to GHG emissions reductions. It does suggest, however, that the task will not be as straightforward.

Strategies for Reducing a Corporate Carbon Footprint

Despite persistent regulatory uncertainty in Canada and elsewhere, some companies have been developing a "sustainability strategy" (including GHG reduction) for several years, under the belief that regulations would eventually be implemented.

The main elements of a corporate carbon reduction strategy are well known. They include the following:

- measurement of GHG emissions
- energy conservation, demand reduction
- fuel switching, e.g. from coal to gas for power generation

- development of a portfolio of renewable energy sources like wind and solar
- carbon sequestration in offset projects like afforestation
- carbon capture and storage

An industry leader like TransAlta, an Alberta-based power producer, is mindful of the challenge and understands the most likely ways to achieve its multiple objectives. The company is well aware that, with 39.2 million tonnes of GHGs released in 2007, they are one of Canada's leading emitters. They also know their energy mix:

- 62 percent coal
- 24 percent gas
- 10 percent hydroelectric
- 4 percent renewables (wind, geothermal)

They are investing in wind power as well as carbon capture and storage. They expect that a pilot carbon capture and storage project will reduce CO_2 emissions by a million tonnes per year. They also point out (politely) that governments are not helping by devising different regulatory frameworks:

> The long-standing "triple E" equation at the heart of our industry is: economic growth = energy growth = emissions. Simply put, this equation must be broken. The first part of the equation requires consumers to reduce demand and be more efficient in their use of energy. The latter requires industry and government to collaborate to more quickly introduce new technologies to drive down emissions....
>
> We are clear that coal-fired generation remains the most economically viable means to deliver electricity reliability at a global level. The challenge becomes significantly reducing emissions of CO_2 to the environment while remaining competitive. We believe that post-combustion CO_2 capture and storage is a highly promising solution while clean coal technologies are being advanced. TransAlta is currently working with suppliers and governments to determine *an appropriate cost-sharing arrangement* for a carbon capture and storage project at our Alberta operations.
>
> We are also working with governments to *encourage a more uniform*—and sustainable—*approach to reducing greenhouse gas*

> *emissions*. During 2007 and early 2008, the governments of Canada, Alberta, Ontario, and Washington state each released their own plan. Harmonization of rules and regulations is essential in order for us to move forward efficiently. (TransAlta Corporation 2008; emphasis added.)

The message here is clear. The company recognizes that it is living in a carbon-constrained world. There is a plan to reduce emissions. However governments offering to share the development costs of new technologies would benefit everyone. Furthermore, governments must harmonize their emission-reduction plans in order to avoid unnecessary costs to industry.

The company is sanguine about the timescale and cost. They note that carbon capture is "still in development" and "will not mature to scale for 10 to 20 years" (TransAlta Corporation 2008). And it will be expensive. The current cost of producing power from all sources is $42 per megawatt hour; the incremental costs to retrofit an existing coal plant are between $40 and $120 per megawatt hour. However, on the plus side they believe that this technology could capture 90 percent of the total CO_2 emitted by a power plant.

It is worth noting that these are only the engineering costs. There are significant legal issues around carbon capture and storage that remain to be explored. Who owns the rights to underground carbon storage capacity? Who would be liable, and over what time frame, for any carbon leakage? Many of these legal questions promise to be knotty. Another example is nuclear power, which is more recently back on the agenda. But virtually no progress has been made on determining the long-term responsibility for the storage and management of high-level nuclear waste since the 1960s expansion of nuclear power. Governments—such as the Ontario provincial government—declare that any new nuclear plants must be built and owned by the private sector. Yet those prospective investors refuse to be held responsible for the long-term storage problem.

Although problems remain, there is some progress in corporate response to climate change. A clear indicator of this is the Carbon Disclosure Project.

Canada and the Carbon Disclosure Project

The Carbon Disclosure Project (CDP) was started in 2001 as a survey of "climate change readiness" of the 500 companies listed in the FT500 Global Index, an index compiled by the *Financial Times* of the 500 largest, publicly listed companies (by capitalisation) in the world. The survey was commissioned by major institutional investors in order to find out whether the big companies in which they invested their resources were prepared to address the potential shareholder value implications of climate change. Table 8-1 summarises the questions asked in the survey. The results may be examined at www.cdproject.net.

Table 8-1 The Carbon Disclosure Project Questions

1. What does climate change risk mean for your company?
2. How do existing GHG regulations affect you?
3. Are your operations affected by physical impacts associated with climate change?
4. What innovations have you adopted in response?
5. Who on your board is responsible for your response?
6. What quantity of the 6 GHGs do you emit?
7. What quantity of the 6 GHGs are associated with the use of your products and your supply chain?
8. What is your emission reduction strategy?
9. What is your strategy for GHG trading in the EU ETS, CDM/JI markets , and other trading platforms?
10. What are the costs of your energy consumption? (primary and electricity)

Source: *Carbon Disclosure Project Report 2007: Canada 200*

In 2006 the survey underwent a major expansion to include groups of companies from regions that were not strongly represented in the FT500Global Index. For example, the Canadian survey targeted the 200 most valuable companies listed on the Toronto Stock Exchange (Conference Board of Canada, 2007). For the *CDP Canada 200 Report*, 97 companies responded in 2009, compared with 78 in 2006. In 2009 the questionnaire was sent to 3,700 companies worldwide.

Canadian investors involved in the CDP include the Caisse de dépôt et placement du Québec and Canada Pension Plan Investment Board.

Table 8-2 shows how quickly the CDP has grown in terms of investor-support and corporate response to the questionnaire.

Table 8-2 Carbon Disclosure Project—Growth and Response Rate

Carbon Disclosure Project	Number of Investors	Assets Under Management	Percentage Reponse to the Questionnaire
2002	35	$4.5 trillion	47
2004	95	$10 trillion	59
2005	155	$20 trillion	71
2006	225	$31 trillion	72
2007	315	$40 trillion	76
2008	385	$57 trillion	85
2009	475	$55 trillion	n/a

Source: *Carbon Disclosure Project Report 2009: Canada 200*

Not only has the project grown in size and response, it has also grown in scope in terms of analysing a company's carbon exposure. For example, in the first round when companies were asked what carbon risks might lurk in their supply chain, even BP—a leader in terms of carbon risk assessment—dismissed the question as impossible to respond to, as in BP's case the supply chain involved more than 3,000 companies. However, by 2007 a number of companies responding to the CDP formed the Supply Chain Leadership Coalition to tackle the problem.

Two generalisations can be drawn from the CDP so far. First, awareness of the carbon risk far outruns actions taken to reduce the risk, although this gap is closing. Second, there is a surprising variety of responses to the risk *within* economic sectors, even major sectors like oil and gas, and insurance. This might imply that, as the impacts of climate change deepen, a significant competitive advantage might accrue to those companies that are best prepared. Assessments of the quality of the responses by the Conference Board of Canada identified 15 Climate Disclosure Leaders "on the basis of their superior levels of transparency" (Conference Board of Canada 2009), as listed in Table 8-3.

Table 8-3 2009 *Canada 200* Climate Disclosure Leaders

High-Carbon Impact Sector	Low-Carbon Impact Sector
Canadian National Railways	Royal Bank of Canada
Enbridge Inc.	Toronto-Dominion Bank
EnCana	Bank of Montreal
Suncor Energy Inc.	BCE Inc.
Bombardier Inc.	Canadian Imperial Bank of Commerce
ARC Energy Trust	
Catalyst Paper	
Gaz Metro LP.	
Emera Inc.	
Penn West Energy Trust	

Source: *Carbon Disclosure Project Report 2009: Canada 200*

Specifically, for several sectors the CDP identified "climate leaders" based on their CDP responses for:

- strategic awareness
- management accountability
- emissions management and reporting
- emissions trading
- programs in place for emissions cuts
- targets established

"High-carbon impact sectors" were classified as any company with a combustion facility with a rated thermal input exceeding 20 MW, and any company from the following sectors:

- automobiles and components
- aerospace and defence
- chemicals
- construction materials
- electric utilities

- energy equipment and services
- oil, gas and consumable fuels
- metals and mining
- paper and forest products
- transportation.

From the survey it was apparent that regulatory uncertainty was seen as the greatest risk by corporate Canada, an even greater risk than the physical threats of climate change.

> Of the 71 respondents that recognize risks from climate change, 59 see regulatory risks as being the areas of greatest risk for their operations (for example, those arising from regulation of GHG reductions) followed by physical risks (such as those from increased severe weather events). (Conference Board of Canada, 2009, page 13)

Table 8-4 shows that response varied greatly between sectors. The visibility of the banks' operations in the public eye means that all the big banks are now aware of their role in climate change even if they are in the low-carbon category, as defined in Table 8-3. They may not emit much in the way of GHG but everyone is aware that the banks' clients include all the major emitters and that bank lending could have a strong influence on clients' behaviour. Likewise, utilities and the retail and consumer sector had a high level of response. Sectoral aggregation means that some important distinctions are obscured. For example, in the transportation sector the two respondents are CN and CP railways, while the non-respondents are airlines. (The dominant response for each sector—"answered," "did not answer" is bolded in Table 8-4.)

The Carbon Disclosure Project provides a very useful awareness-raising opportunity for the Canadian economy, especially given the tardiness of senior governments to increase the pace of response to the climate change challenge. We can expect that the response rate to the CDP will climb rapidly once GHG emissions are capped across Canada. In the meantime, the carbon economy is alive and well in the European Union, as well as in parts of the United States and

Table 8-4 Company Response to the *CDP Canada 2009* by Sector

Sector	Answered the Questionnaire	Did Not Answer the Questionnaire
Chemicals and Pharmaceuticals	**4**	2
Construction and Building Products	**3**	1
Financial Services	**15**	9
Hospitality, Leisure, and Business Services	10	**14**
Manufacturing	1	**2**
Mining	10	**16**
Oil and Gas	14	**22**
Publishing	–	**2**
Raw Materials, Paper, and Packaging	**4**	2
Retail and Consumer	**13**	7
Technology, Media, and Telecoms	**9**	4
Transport and Logistics	2	**5**
Utilities	**9**	3

Source: *Carbon Disclosure Project Report 2009: Canada 200..*

Australia. The opportunities that will soon present themselves to Canadian businesses will be introduced in the remaining sections of this chapter.

Trading Opportunities in the Carbon Economy

Unlike a carbon tax, the cap-and-trade approach explicitly creates new markets. Given that we have seen in the previous section that companies within the same economic sector treat the carbon risk very differently, the competitive advantage will go to those companies that assign the correct risk assessment to their position in a carbon-constrained world.

Compared with the European Union, the other capped Kyoto signatories have been unable to move quickly, as their governments have been reluctant to embrace the opportunities provided by a concerted drive to reduce GHGs. Of course there are risks lurking within these opportunities. We've seen the price plunge in Phase One of the EU ETS. There are several features of the new carbon economy that make it particularly risky.

First, it is created by regulation and so depends on the commitment of governments to keep to a path that may not always be popular. Governments may slacken off or toughen up, depending on their political stomach. Second, it will initially depend on government funding to establish the trading framework, such as a transaction log; this of course takes time to set up. We have already seen that serious underfunding led to unfortunate delays in setting up the Clean Development Mechanism for the Kyoto Protocol.

Third, the carbon economy is quintessentially *global*. Credits from capturing methane from pig manure in Brazil can be used by Canada to help meet its GHG reduction goal under the Kyoto Protocol. Life does not get more global than that. By their nature international transactions add an element of risk in a way that domestic transactions do not.

Fourth, the idea of a carbon economy is entirely new. Players have little experience, and many have none at all. This means that mistakes will be made, and surprises could be costly. Also, we have no control over how fast or slow the physical process of climate change will take place, and this is the process that is driving the market. Climate processes could suddenly speed up (recall those positive feedbacks we encountered earlier) and force governments to take more drastic steps, such as increasing the fines companies pay for non-compliance, or tightening caps on emissions.

Fifth, the carbon-constrained world is new territory where there is little agreement on who is responsible for what—either morally or legally. As Canada's permafrost melts (in response to the *global* burden of GHGs in the atmosphere), and begins to emit large quantities of methane, the question arises: should that extra GHG burden be placed squarely on Canada's Kyoto balance sheet, or should all emitters around the world be billed for their "share" of those emissions?

Trading opportunities arise in different segments of the carbon market. There is an "allowance market" in which capped installations—such as a power plant—reduce emissions in order to meet the cap. If they exceed their target reduction they may sell their surplus carbon credits to other emitters who have either failed to meet a cap of their own, or are simply trading in carbon credits on a speculative basis. For some traders carbon has already become "just another commodity" like corn or pork bellies or orange juice. Then there is the "project market" where entities which are not capped under Kyoto develop projects that will reduce emissions to gain credits within the CDM or Joint Implementation frameworks (see pages 80–81). The entity in the host country that owns these credits can then sell them to a "carbon credit aggregator" who, in turn, sells them to a company or country that needs to meet a compliance target.

Both the allowance market and the project market developed under the "compliance market," put in place by the Kyoto Protocol. There is a parallel activity known as the "voluntary market." Here individuals and companies reduce GHG emissions for a combination of altruistic and marketing reasons. For example, airlines offer environmentally concerned passengers the opportunity to "offset" the CO_2 emissions associated with their flight by planting enough trees to absorb the CO_2. Likewise a company or an event may advertise itself as "carbon neutral" to attract customers. Conferences on an environmental topic might also advertise themselves as carbon neutral. This means that they have minimized the GHGs they emit and have offset what remains by planting trees or investing in renewable energy projects. The earliest voluntary market for trading GHG reduction credits was established in Chicago in 2003 as the Chicago Climate Exchange (CCX) in anticipation of forthcoming regulations on emissions in the US. In 2008 the CCX established the Montréal Climate Exchange within the Montréal Exchange.

The "fundamentals" of a compliance market like the EU ETS rest on the emission caps set by the regulatory authority, plus other factors such as the associated fines for non-compliance, the integrity of the reporting procedure, and the promise of continuity. Other important fundamentals include the weather and the price of the various sources of energy. For example, dry weather may reduce available hydroelectric power and hence drive up demand for other energy sources, typically coal and gas. So with more GHGs emitted, dry weather will indirectly drive up the price of carbon reduction credits, especially in countries like Sweden and Spain where hydroelectric power is a significant part of the mix. Temperature is another fundamental driver of the compliance market, given that hot summers and cold winters drive up demand for space cooling or heating.

In addition to these fundamentals, the carbon market—like all markets—is also driven by "sentiment," or anticipation of market direction. Phase One of the EU ETS provided a perfect example of the impact of sentiment on prices. The market opened in January 2005 and the price advanced steadily towards 30 euros per tonne of CO_2, which was the fine set for non-compliance. Almost everyone who was following this nascent market (including myself) was confident that the price reflected the realities of demand and supply, in other words, the fundamentals. However, as the first truing-up moment for 2005 emissions approached in April 2006, data were leaked which suggested that the market, overall, was going to be long. The price plummeted in 48 hours to about eight euros. There were still some buyers in the market who were short, and finally six of the twenty-six EU member states were short; but the market overall was significantly oversupplied with allowances. Trading for Phase One carbon credits was all but dead.

The EU ETS experience should provide useful material for establishing a carbon trading system in Canada. Unfortunately, the degree of fragmentation in the Canadian systems will drive up the price of carbon, as opportunities for trading will be more restricted. However, eventually the carbon market will mature both Canada-wide and globally, and it will assume all the features of other commodity markets—with swaps, options, and futures.

One small example of a deal that did not go through should illustrate the potential. As early as 2002 Hydro-Québec and the city of Montreal tried to arrange a "weather swap" based on the quantity of snow falling over the winter. This was a perfect "natural swap": a lot of snow would leave the city with a big bill for snow clearance, but would also leave Hydro-Québec with a healthy reserve for generating electricity in the spring. Each hoped to be able to offset part of their risk by a schedule of payments based on the actual snowfall. In the end the deal did not go through because of a lack of agreement on the baseline for data.

In a mature carbon market a company exposed to risks from the price of carbon, the variability of the weather, and the price of various sources of energy should manage all three elements of risk by developing appropriate hedging strategies. One of the great attractions for investors—even if they are not a capped entity—is that the risks in the carbon market are not closely correlated with the other traded stocks and shares, and hence offer the investor a very useful opportunity for diversification.

Opportunities for the Legal and Financial Professions

The impacts of climate change and our attempts to mitigate the risk while adapting to changing circumstances on the ground will be so pervasive that it is hard to think of any sector of the Canadian economy that will not be affected. However a glance at Table 8-4 suggests that a good portion of the 200 largest companies on the Toronto Stock Exchange are either unaware of the risk or believe that it is manageable for their company. Keep in mind that only 44 percent of the companies replied to the survey.

It is easy to see that climate change is of immediate consequence for forestry, agriculture, transportation, construction—indeed for almost any outdoor activity. Perhaps it is less obvious that many largely indoor activities will also be transformed by climate change. We will be living in a different kind of world once the physical impacts of climate change become more widespread and once the price of carbon permeates the global economy. Given that we are beset with a long list of questions concerning responsibility for climate change and for devising an appropriate response, the legal profession will be in the thick of it, both in Canada and globally.

Climate change raises legal issues, from the global/international to the local. At the global scale, the Alliance of Small Island States has sought legal redress for their plight. There is no question that they have suffered a loss for which they are not responsible. But who is? Within the realm of Canadian constitutional law, climate change is a clear challenge: environmental issues are a shared federal/provincial responsibility. There is no obvious way of dividing up the responsibility.

Likewise, many problems remain at the municipal level. Municipalities are responsible for the health and welfare of people living within their jurisdictions, yet given the lack of funds—not to mention the lack of political authority—dealing comprehensively with climate change is no simple challenge. There will be multiple repercussions on municipal operations from traffic management and road repairs, to flood management and response to heat alerts.

New challenges will also be raised for commercial law. The point has already been made that as a commodity, carbon entails new elements of risk. For example the credits for a CDM project could disappear overnight if the executive board declared them to be invalid. Who would cover the counterparty risk in such situations? The potentially ephemeral nature of carbon credits has already driven one carbon aggregator to delist from the London Stock Exchange and seek protection from bankruptcy.

Similarly, engineers are developing new technologies that are so unlike previous innovations that current laws may be difficult to apply. For example, under the evolving technology for carbon capture and storage, CO_2 will be pumped and stored in fissures underground. However, responsibility for any subsequent leakage will be difficult to assign—most law on "underground rights" relates to removing material (minerals or water), not adding it.

Within the financial services sector it was the property and casualty insurance industry that was the first to find itself embroiled in consequences of climate change, mainly due to record payouts for extreme weather events such as hurricanes in the US and winter storms in northern Europe. Hurricane Andrew provided the wake-up call, with an insured payout of approximately $20 billion (in current US dollars) even though the hurricane itself missed any major urban centre. The event produced a dozen bankruptcies in the insurance industry and shook the global reinsurance system. Subsequent research by insurers revealed that had the local building code been followed rigorously, losses would have been only 20 percent of what they actually were. Armed with this information, the industry began a new proactive approach in preparing for an increase in losses from extreme weather events. The Insurance Bureau of Canada set up the Institute for Catastrophic Loss Reduction to develop a response, beginning with raising awareness of the issue across Canada. The Insurance Bureau, together with the University of Toronto and Environment Canada, put together a team to assess how Canada would respond to extreme weather events based on past responses (Brun et al. 1997)

The insurance industry was perhaps the most prepared for a quick response, having been previously embroiled in costly environmental problems generally known as "asbestos and environmental." Arising from the 1970s onwards, the problems included pollution from leaking landfills, as well as asbestosis and mesothelioma caused by the widespread use of asbestos in industry, shipping, and buildings. Some prominent insurance companies remained in denial over the exposure to asbestos for twenty years; finally, however, they had to respond. As losses from climate change could easily dwarf the billions paid out for asbestos, many insurance companies in Canada and Europe moved quickly to assess their exposure and prepare themselves and their customers for a long term adjustment.

This adjustment included both climate change itself and the operational changes the industry would have to make.

Initially, banks in Canada were much slower than the insurers to see that they too faced problems. This may have been due to the presence of CEOs from high-carbon companies on their boards—oil and gas, mining, chemicals, and forest products. Up to the mid-1990s opinion leaders from these sectors were for the most part skeptical, and therefore opposed to any government action on GHG reduction. However, the banks—like the insurers—had considerable experience with environmental problems such as contaminated land, which also proved to be intractable and expensive. Eventually opinion shifted from dismissive to proactive. Both commercial banks and investment banks now show an awareness that climate change is an issue for them and their clients, especially clients on the high-carbon side of the table.

In the public eye, perhaps, the accounting profession faces fewer concerns. But the issues facing this profession are in fact much broader than the climate change issue; these relate to the notion of environmental performance across a range of indicators. These concerns arise from a recognition that

> sufficient research has been done to establish a clear relationship between superior environmental and social performance and a company's long-term value creation—perhaps it's to do with reputation and trust-building, perhaps with superior risk management, efficiency, or innovation. (Willis, in Labatt and White 2002)

The ambition to capture this dimension of management led to the establishment of the Global Reporting Initiative in 1997. In 2002 this organization produced a set of guidelines to advise the profession on how to account for environmental liabilities. Adopters of this initiative were able to stay ahead of the wave of legislative reform that—in Europe and the US—has required a fuller accounting of environmental problems now appearing on the balance sheet as "contingent liabilities."

In Canada, "accounting for climate change" has become a central concern of the Canadian Institute of Chartered

Accountants, which has issued guidelines as to how this should be done in financial statements and in the *Management's Discussion and Analysis Report* required from public companies. (See http://www.cica.ca.)

Opportunities for Clean Low-Carbon Technology

A cap-and-trade system for GHG emission reductions will drive innovation. As long as fossil fuel energy was cheap, the payback time for renewable energy could be long. Five years ago friends of mine in England calculated that it would take 40 years for their passive solar heating system, newly installed on their roof, to pay for itself, and it would probably need to be replaced before then. Other regulatory incentives—like renewable energy portfolio requirements and preferential feed-in tariffs to the grid—add to the pressure to innovate.

The most visible result of this pressure is the boom in the "cleantech" sector. Seventy-five percent of the venture capital going into cleantech in North America is for innovations in energy such as demand-response software for utilities, wind turbine technology, and solar photovoltaics. Water is also an important component of the cleantech sector, especially for infrastructure management (such as leak detection and repair), desalination, and reverse osmosis. Other targets for cleantech innovation are transportation systems and solid waste management.

GHG reduction in a compliance market is driving the development of software specially designed to "manage carbon" and provide the audit trail required by regulators. Carbonetworks is a BC company developing this new business opportunity. Clearly, transportation systems will require a transformation to provide net benefits in a carbon-constrained world, especially as this sector is still growing even in countries like Canada and the US where car ownership is near saturation. In the hunt for a clean-fuel automobile, the fuel cell has already been put on hold after demonstrating considerable promise. Electric cars are becoming more common on the roads, as are refuelling points. Toronto-based Zenn Motor Company has developed an "ultracapacitor" approach that transfers energy to the car in a matter of minutes.

In order to bring cleantech to market, demands are placed on a new type of investor that specializes in green technology.

Prices, Behaviour, and Markets

Putting a price on carbon by setting up a GHG cap-and-trade system is an essential part of adjusting to living in a carbon-constrained world. Because we do not know in advance how individual behaviour will change in response to a carbon tax, we cannot rely on such an approach—on its own—to bring about necessary changes. In spite of many studies we cannot predict the elasticity of an individual's demand for personal car travel. Nor do we know if a modification in behaviour will be long term. London's congestion charge did have an immediate impact on the modal split, moving 20 percent of commuters from car to bus in the first twelve months. But it is not clear how long the change will last.

We do know that a cap-and-trade system for GHGs, plus a renewable energy portfolio requirement and attractive feed-in tariffs, will achieve results. We can see this happening in Europe. Unfortunately for innovators in the Canadian business community, the timidity of the senior levels of government have not provided a comparable level of incentive in Canada.

Chapter 9

Canada's Record on Greenhouse Gas Emissions

Canada has produced a plethora of plans, strategies, and consultations since 1990.... A comprehensive study by the Pembina Institute concluded that by the year 2000, only one-third of these actions had been taken, almost all of which involved soft measures such as voluntary initiatives, education, or research.... Canada's record in responding to climate change, both internationally and domestically, is dismal.

David Boyd, *Unnatural Law*, 2003, 86–87, 92

On September 28, 2006, the Commissioner of the Environment and Sustainable Development released her 2006 Report on Climate Change. *The* Report *described that even though the federal government had announced billions of dollars in funding since 1992 towards meeting commitments to address GHG emissions, as of 2004 Canada's GHG emissions were 26.6 percent above 1990 levels.*

Canada's Fourth National Report on Climate Change, 2006

Although Canada is still officially a party to the Kyoto Protocol, it will follow an alternative reductions schedule (postponing reaching its Kyoto target to beyond 2020). Emissions trading is still under consideration, but with limited linking to an international carbon market.

Karan Capoor and Philippe Ambrosi,
The State and Trends of the Carbon Market, 2008, 49

Canada's Trajectory on GHG Emissions and Policy

From the government's own copious records it is clear that Canada has yet to take a serious step along the long road to meeting its Kyoto commitment, let alone the even longer road to stabilizing GHG emissions worldwide. How do we compare with other industrial nations that agreed to caps on their emissions under Kyoto?

Table 9-1 Carbon Dioxide Emissions from Selected Countries, Percentage Change 1994–2004

Country	Ranking in Global CO_2 Emissions, 2004	Percentage Change 1994–2004
United States	1	13
Russia	3	0
Japan	4	16
Germany	6	–1
Canada	7	19
United Kingdom	8	2
Italy	10	21
France	12	13
Australia	14	38
Spain	18	55
Poland	21	–10
Netherlands	22	21
Belgium	27	16
Austria	41	23
Portugal	43	38
Finland	45	7
Sweden	46	0
Hungary	48	–4
Denmark	49	–13
Norway	53	43
Bulgaria	54	–7
Switzerland	55	6
Ireland	56	46
New Zealand	59	22

Source: US Energy Information Administration 2007. Available from http://www.eia.doe.gov/emeu/international/contents.html.

We can see that Canada is not alone in allowing CO_2 emissions to rise. The data in the table predate the entry into force of the Kyoto Protocol (February 2005). However, post-Kyoto emissions broadly follow these earlier patterns. In other words, in terms of actual emissions, we do not yet see any definite impact of the Protocol. The earlier trends include national policies favouring renewable energy, such as wind (Denmark), solar (Germany), and biomass (Sweden). The rise of emissions in the UK was very modest as the country closed the last of its coal mines and switched to North Sea gas. Falling and flat emissions trajectories in Hungary, Poland, and Russia can be attributed to the collapse of heavy industry following the fall of Communism. A continued rise of emissions in Spain and Portugal was expected (already allocated within the European Union "bubble") as their economies continued to grow and modernize after joining the EU. Steady increases in the US and Canada have reflected the business-as-usual approach. Their economies grew, but no emissions policies were put in place.

None of this excuses the policy vacuum that existed, and continues to exist, in Canada.

We have previously noted how during the long period of Liberal federal government (1993–2006) climate change policy vacillated. For the first Conference of the Parties to the Kyoto Protocol (in Berlin in 1995), Canada was represented by Sheila Copps, an environment minister with a keen understanding of the seriousness of climate change. For those who shared this viewpoint it seemed like an excellent start. But after the conference, Prime Minister Jean Chrétien switched Copps to a new portfolio, Multiculturalism. While other potentially competent ministers were later appointed to the post, the prime minister's office was clearly not in favour of any commitment to a proactive stance. What emerged instead was the *de facto* "policy" of producing proposals, while doing very little to prepare Canada to reduce emissions.

South of the border there was little progress during the Clinton years, apparently due to lack of support in Congress. With the election of George W. Bush in 2000, whatever policy doors might have been open in Washington were swiftly closed.

We saw how Chrétien announced at the 2002 UN conference in Johannesburg that his government would ratify the Protocol by a vote in Parliament. While Canada did go on to ratify the Protocol, there was no visible change to policy. The subsequent election of a minority Conservative government in Ottawa under Stephen Harper meant that climate change slipped even lower on the agenda, if such a thing were possible. There were more proposals—including the paper hopefully titled *Turning the Corner*—and the government continued to send those voluminous reports to the UNFCCC.

Meanwhile, consensus about the risk of climate change continued to build. In May 2006, a group of Canadian economists sent Harper an open letter:

> The increasing consensus over the last fifteen years in the natural sciences community, including the conclusions of the highly respected IPCC, cannot be dismissed.... Given the long lead-times for some capital replacement (e.g. electricity generation equipment), public policy signals that business-as-usual GHG emissions may no longer continue need to be put in place immediately. (Open letter to the prime minister of Canada on climate change policy from Canadian economists.)

This plea, like others before it, fell on deaf ears. As "environment" is a shared federal-provincial responsibility, the federal government was prepared to see the provinces pursue separate paths and then wait to compare each of them to its own (proposed) plan. Those provincial plans that were deemed "equivalent" to Ottawa's would stand; weaker plans would have to be beefed up. The federal proposal consisted at that time of intensity targets for major emitters, payments into a technology fund if the targets were not met, and offset projects that included carbon capture and storage, landfill gas capture, and fifteen other activities. Among the provinces, only Alberta had something that resembled the federal proposal. Four provinces opted for a cap-and-trade agreement with regional groups in the US. British Columbia signed an international treaty to trade carbon with the EU. Some provinces remained uncommitted to anything. Canada was on the way to building a regulatory jigsaw puzzle that would be a nightmare for any company expected to comply.

As 2008 unfolded it became increasingly clear that this apathetic approach was becoming increasingly untenable. It was an election year in the US; the Bush era was finally coming to a close. By the time of the Democratic nomination the three remaining contenders for the presidency had all firmly endorsed a federal cap-and-trade system for managing GHG emissions. Whoever won, the new president would hope to establish such a system as an urgent priority. Would federal Canada join? Half of the provinces of Canada had either already joined American state-led initiatives (such as the Western Climate Initiative or the Regional Greenhouse Gas Initiative) or held observer status.

Soon, the question shifted to this one: "Would federal Canada have any choice?"

From the early days of the EU Emissions Trading Scheme, there were dark mutterings within the EU about placing tariffs on imports coming from countries without comparable carbon caps in place—in other words, the rest of the world. The US was the first to note that this would contravene World Trade Organization rules. However, even when the Bush administration was still in charge we heard suggestions that China, for example, should expect its exports to the US to bear a carbon tariff, if the US moved to cap GHGs.

In the closing months of the presidential campaign, as a cap-and-trade system became ever more likely, there was discussion about such tariffs being widely applied to every country's exports to the US, including—naturally—Canada's. (In other words, 80 percent of all Canada's exports.) As the oil price fell and the recession deepened, this dilemma intensified. The problem became particularly acute for Canada's oil sands projects under development in Alberta and Saskatchewan. Investors had already started to pull out in response to the fall in the oil price, the recession, and the credit crunch. Furthermore, California put legislation in place to target Canada's "dirty oil sands," while the new Obama administration was promising something similar.

Suddenly, "climate change in Canada" was no longer about pine beetles and polar bears: it was about the lifeblood of the Canadian economy.

Minister of the Environment Jim Prentice was speedily dispatched to Washington to obtain some clarity (and perhaps comfort) about the American linkage of climate change to tariffs on international trade. In so doing he provided *The Globe and Mail* (2 March 2009) with a headline opportunity that comes rarely in a journalist's lifetime:

> Prentice hits Washington, with cap (and trade) in hand.

The word "comparable" suddenly became a double-edged sword. It was no longer a matter of the provinces producing legislation that was comparable to the federal proposals. Now Ottawa would have to enact comparable legislation to whatever came out of Washington to avoid trade penalties on the carbon content of its exports to the US, beginning with petroleum products from the oil sands.

It is possible that this about-face in Ottawa was made not only in response to anticipated pressure from Washington. Another factor may be the realization that although emission intensity regimes (as proposed by Ottawa and already enacted by Alberta) are designed to provide room for economic growth, they become a trap in a recession, given that capped emitters still must meet their intensity target whatever their level of output. In a recession, intensity targets provide an additional burden for companies already facing reduced demand and tight credit. At present no one is predicting an imminent end to the downturn.

The reasons for Canada's climate change policy predicament are also undergoing re-evaluation. There now appears to be a strong desire to blame everything on Bush's lack of leadership:

> It was hard, Mr. Harper said yesterday, to make progress in Canada when no willing partner existed in the US.
>
> It turns out by this twisting of history that Canada's terrible record was the fault of Mr. Obama's predecessor, George Bush.

> To suggest that Canada's failure [on climate change] was because the Americans were an unwilling partner is historical revisionism of a brazen kind. (Jeffrey Simpson, *The Globe and Mail*, 20 February 2009)

Despite the finger pointing and uncertainty about the economic and environmental future, a new and relatively simple understanding of the Canadian situation emerged. It was summarized in just four words in a *Globe and Mail* headline: "Hard caps, free trade." The article continued somewhat optimistically:

> Canadian officials must act on the assumption that Congress will pass legislation of this kind before the United Nations Climate Change Conference in December in Copenhagen, which will aim to replace the Kyoto Protocol.
>
> Canada is now likely to be ready this year with a respectable cap-and-trade plan. (*The Globe and Mail*, 9 April 2009)

International Opinion on Climate Change

While the Canadian government was drowning in a policy morass of its own making, much of the rest of the world has been moving ahead.

The European Union

The climate change news from the European Union is mixed. Critics of the idea of carbon markets in general, and of the EU ETS in particular, have been quick to seize on the difficulties experienced by the world's first compliance market. In this market, capped facilities must comply with a given cap, either by making the necessary reductions in emissions or by buying carbon credits to meet their deficit. This choice is identified succinctly as "make or buy." There have been difficulties, but even so the EU now has more than four years of carbon trading to its credit, while the US is still struggling to pass a climate bill through Congress. Canada is not even at the point of introducing a bill.

We noted earlier that some of the weaknesses experienced in Phase One of the EU ETS (2005–2007) resulted from deficiencies in the design of the scheme, arising from compromises intended to ensure buy-in from the business community. Also the designers wanted a start date of 2005 to provide an opportunity to learn from three years' experience before Kyoto's First Compliance Period began. Thus:

- All the emission allowances were freely distributed, rather than auctioned.
- The baseline for emissions was not independently verified until the scheme was under way.
- The timeline for Phase One (three years) was too short to match the typical business investment cycle.
- Credits from Phase One could not be banked for use in Phase Two, and hence became worthless at the end of the first phase.

However, despite some setbacks, the EU carbon market is in place. The scope of the market is steadily growing; it is fully expected that the market will be well established at the end of the First Compliance Period in 2012, whether other countries join in or not. For carbon market designers and managers in North America and elsewhere, there are many valuable lessons to be learned from the EU experience.

In Britain, as in other EU countries, there was much skepticism about the scheme; and some skepticism remains. Yet with the 2006 publication of the British government's *Stern Review* (by the World Bank's former chief economist, Nicholas Stern), we clearly saw that the likely cost of inaction *vastly* outweighed the costs of reducing GHG emissions; a wait-and-see approach was highly imprudent. In 2008 the British Parliament passed their Climate Change Act, which committed the country to a steady reduction in GHGs, regardless of Kyoto or the EU ETS.

The EU's Environmental Commission produced an ancillary directive in January 2008 requiring each member state to meet renewable energy targets. Britain, which currently has only 2 percent in renewables, has been set a target of 15 percent by 2020. Sweden, currently at 37 percent, has been targeted at 49.

The United States

As in Canada, the lack of federal leadership did not deter lower levels of government, businesses, and volunteer groups from taking their own GHG reduction initiatives. One of the earliest and most significant interventions was the Chicago Climate Exchange (CCX), established by 40 corporations. This began trading in September 2003. Although the CCX is a voluntary organization, its members make a legally binding agreement to reduce their emissions. Membership has grown to over 400 and now includes participants from Canada, Chile, China, India, and Mexico.

Initiatives at the state level have already been introduced. Of these the Western Climate Initiative is probably the most ambitious: it proposes to establish caps broadly across several sectors of the economy beginning in 2010. British Columbia, Manitoba, Ontario, and Quebec have joined this plan. The only state-level group that is already in operation is the Regional Greenhouse Gas Initiative (RGGI, known as "Reggie") in the northeast. As a sign of how the thinking on carbon markets has evolved since the European carbon market, in RGGI *all* of the emission allowances are auctioned. However, emission caps are confined to the electrical utility sector. Ontario and Quebec have observer status in RGGI.

It is widely assumed that these state-level initiatives will soon be subsumed within a federal cap-and-trade structure. It is also expected that any reductions made under the regional schemes will be awarded "credit for early action" in a federal market.

We have already mentioned the progress toward the establishment of an American federal carbon market. The Waxman-Markey bill was passed by the House of Representatives and awaits the Senate vote. It is widely anticipated that, despite the opposition of Republican senators and Democrats from coal-producing states, a climate change bill (in some form) will be passed in 2009. As a powerful backup to action by Congress, a recent Supreme Court ruling has determined that GHGs are a "pollutant" under the terms of the Clean Air Act, and thereby subject to regulation by the Environmental Protection Agency (EPA). This implies that if Congress fails to pass a bill then a carbon market could be established by the EPA.

Either way, it seems probable that momentum will gather for the establishment of an American market in GHG emission reduction credits before the end of Kyoto's First Compliance Period in 2012. That market will certainly include rules for tariffs based on the carbon content of all imported goods—beginning with imports from countries that signed the Kyoto Protocol and *failed to set up rules for implementation*. In this target group Canada is the most prominent member. The American train for carbon markets is leaving the station. Does Canada still have enough time to get on board?

Countries without GHG emission caps

Behind all the debates about "Whose carbon is it anyway?" is the historical legacy of "development"—the process by which today's richest countries became so rich in material goods. This process was literally fuelled by burning fossil fuels in the form of coal, oil, and natural gas. The countries that did not accept emission caps under the Kyoto Protocol hope to follow that same development path. Indeed, the richer countries continue to follow it themselves.

The global challenge is to find a means to achieve development without fossil fuels. Everybody understands this. The question is how to devise a different incentive structure to accelerate the transition to a low-carbon economy. Countries at an earlier stage of this development path expect the countries that have already benefited from it to show leadership. This is hardly unreasonable—in fact, this was largely taken for granted when the Kyoto Protocol was signed. The richer countries agreed to reduce their emissions, while the poorer countries (poorer in *per capita* income terms) signed without accepting caps. President Bush and Prime Minister Howard withdrew their countries from the Protocol because they found this distinction unreasonable. History will judge them on this. Both men have since departed the political scene.

The issue has always been a matter of *how much* the richer capped countries would put on the table to make it acceptable for the uncapped countries to reduce their GHG emissions. We need to make urgent progress on this issue. Many of the uncapped countries have been ready to move for some time. China was the first country in the world, rich or poor, to produce a climate action plan following the Earth Summit at Rio de Janeiro in 1992. Among the other uncapped countries, Mexico, Brazil, and Argentina have all signalled that they expect to assume caps of some kind in order to develop a global approach to the challenge of climate change.

The Challenge for Canada

It is widely recognized that Canada's record on GHG emissions, and even on policy development, is extremely sad. This has had consequences for global progress on emission reduction and it has also tarnished Canada's environmental reputation. It has in addition cost Canadian business significant opportunities. The two "bad boys" of the Kyoto Protocol—Australia and the US—have either returned to the fold, or are about to do so. The other capped countries that avoided EU-style initiatives—Canada, Japan, and New Zealand—must now decide how to make up for lost time.

The next chapter outlines developments in the global carbon market, which Canada must soon enter. The final two chapters sketch the risks and opportunities that await us in a carbon-constrained world.

Chapter 10

The Emergence of a Global Carbon Economy

President Barack Obama has repeatedly stated his belief that tackling climate change with a cap-and-trade system will nudge the US toward economic recovery. "The whole rhetoric that tackling climate change will cost jobs is six or seven years old," said University of Victoria climatologist Andrew Weaver. "Most governments have moved beyond that."

Saskatchewan reneges on climate change,
The Globe and Mail, 23 April 2009

The global economic slowdown continues to take its toll on the carbon market with a series of redundancies and consolidations taking place across all sectors of the market.... But the newly formed Bank of America Merrill Lynch has announced it has no plans to "reduce emphasis in the carbon market."

Economic downturn hits carbon jobs,
Carbon Finance 6, no. 2, February 2009

The Steps Ahead for Canada

Sooner or later Canada will establish a cap-and-trade market for GHGs. That much is practically certain: the US will do so and Canada will follow, if only to preserve its most important trade relationship. While perhaps not the most altruistic of reasons, it will nevertheless be the most compelling.

Some politicians and voters may fear that reducing GHG emissions will cost jobs in Canada. As Andrew Weaver and others have observed, that kind of thinking is several years out of date. Even if there are still concerns over job losses, any such estimates must recognize that these losses would pale beside the potential losses from being squeezed out of US markets for failing to follow America.

Stephen Harper belatedly proposed setting up a carbon market jointly with the US. The offer was politely ignored by President Obama and an undefined "dialogue on clean energy" was offered as a replacement. The fact is that the new government in Washington will have a tough enough battle to pass a climate change bill without trying to do so at the same time as building a carbon market with Canada. As outlined in the previous chapter, if the climate bill fails to pass the Senate then the president can fall back on the Supreme Court ruling and leave it to the Environmental Protection Agency (EPA) to design the carbon market, beginning with vehicle emission standards. The senators who currently oppose the passage of the bill must be aware of this option and will eventually accept the reality that a bill with (their) various conditions included is better than relinquishing control of the climate change response to the EPA.

In Canada, we should assume that, one way or another, the US is likely to be living under a GHG cap-and-trade system by the end of the Kyoto First Compliance Period in 2012. The reality of international trade (if for no other reason) is such that Canada has no choice but to follow a parallel process at the federal level. The days of leaving the provinces to follow different initiatives (or, in some cases, none at all) and then assessing them for "equivalence" with Ottawa's latest proposal must come to an end.

This situation might seem straightforward. Is there any alternative? One might not think so. But the situation in Canada is endlessly complex and the road ahead may seem a little less clear to some of the parties involved. Consider for example the following statement that emerged from a recent meeting of the Canadian Council of Ministers of the Environment:

> Last month's meeting [January 2009] of the Canadian Council of Ministers of the Environment (CCME) in Whitehorse noted that "climate change has had significant impacts on northerners' way of life" and "called for an implementation plan and timelines for the proposal to be brought back to CCME for consideration at their next meeting." (CCME, http://www.ccme.ca)

This is hardly a clarion call for action.

The CCME is composed of all environment ministers from the federal, provincial, and territorial governments; they are certainly aware of the climate change bill progressing through US Congress. Members should therefore be well aware of exactly what is needed to develop regulations that would be at least "equivalent" to what emerges south of the border. For Ottawa to fail to devise a course of action that could be supported by Parliament and the provinces would be an extreme failure of leadership. Alberta will need to replace its "intensity approach" with cap-and-trade regulation in order to be "equivalent." If we are still in a recession at the time when these decisions are taken, then Alberta might be quite willing to escape from the "intensity trap" in which it finds itself.

Canada has no choice but to develop an approach to GHG mitigation that is equivalent to whatever emerges from Washington in the next few months. That much is obvious. But Canada must prepare for even more than this. Over the years in which the leaders of Canada, the US, and other countries have ignored climate change, the "carbon economy" has become a commercial reality. This is the world to which Canada must quickly adjust. The remainder of this chapter will assess the current state of the global carbon market and the challenges that market poses.

The Carbon Economy to Date

The carbon market has grown rapidly since 2004, although we have seen how it is dominated by the European Union Emission Trading Scheme (EU ETS) and the Clean Development Mechanism (CDM) offsets that feed into it. Despite setbacks, the EU ETS became the first functioning compliance market for carbon that the world has ever seen. As the name suggests, players in a "compliance market" have no choice but to participate. There is also a parallel "voluntary market," in which individuals and organizations set their own reduction targets—such as the Chicago Climate Exchange, introduced in the previous chapter. Participants in both types of carbon market may make use of "offset projects," where their GHG emissions are offset by some compensatory action in a jurisdiction (a country, or a sector of the economy) without capped emissions. Offsets may take a number of forms. The majority occur either in industry, forestry, agriculture, or energy. This chapter will consider examples of each type.

A key factor is whether an offset project is domestic or offshore. The distinction is important for the integrity of GHG emission reductions in a capped jurisdiction and also for the political and financial dimensions of the employment benefit of investment in offset projects. The integrity issue revolves around the allowable percentage of an entity's emissions reduction that can be made up by offshore offsets. Clearly a very high percentage would significantly reduce the need for actual GHG emission reduction domestically. We saw above how in the early days of the EU ETS, Ireland proposed allowing 50 percent in offsets. Subsequently the EU Environmental Commission brought in a ceiling for member states of 10 percent, with a recommended range of 5 to 10 percent.

There is also a potential employment benefit from encouraging domestic rather than offshore offsets. For example, under Alberta's Climate Change and Emissions Management Act *all* offsets against its intensity targets must be located within the province. This provides some local employment benefit. However local offsets could be more expensive than overseas offsets, and so drive up the cost of compliance.

Transaction costs in the carbon market, as in other markets, are a function of volume and hence liquidity, as well as the transparency of prices. As the market matures, fewer transactions will be made "over the counter"—that is, bilateral and private—compared with transactions traded visibly on an exchange. We have already begun to see evidence of this:

> Brokers remarked that there were higher volumes of on-screen trading in RGGI allowances than in the over-the-counter market, a trend that began two weeks ago. (RGGI allowances set record in third auction, *Carbon Market North America*, Point Carbon, 20 March 2009)

Thus the carbon market is driven by the need for capped entities, such as electrical utilities, to reduce their GHG emissions. How much this will cost will depend on the regulations governing the market. Possible regulations include the following issues: the caps in relation to current emissions, the extent to which allowances are auctioned, the extent to which allowances may be banked from one year to the next, the limits on the use of offsets (and whether these offsets must be local or can be sourced globally), the protocols that govern how emissions will be measured, and so on. For example, protocols for measuring the impacts of offsets (their emission effectiveness) may be based on those already developed for the CDM. Alternatively they could follow standards set by the International Organization for Standardization (ISO), or standards developed by a third party such as The Gold Standard. The greater the buyers' confidence in the integrity of offsets, the greater the value of the offsets on the market.

In the early days of the CDM the most popular offsets were those provided by the reduction of the use of nitrous oxide (N_2O) and sulphur hexafluoride (SF_6) in manufacturing processes in factories. These gases have a very high global warming potential and so yield huge quantities of reduction credits. However, they do little to advance the development goals of the CDM. Offsets may also be gained from renewable energy projects such as wind, solar, tidal, small hydro, and biomass. Like the industrial projects, the impacts of these activities are relatively easy to measure, although it may be difficult to agree on what type of power they are displacing—gas, coal, lignite, nuclear, large hydro? It is more difficult to measure the impacts of forestry and agriculture, given that these depend on biological processes rather than mechanical systems.

It is difficult to know exactly how much carbon a forest is sequestering as it grows, as it is simultaneously in the process of releasing carbon as trees rot, are chopped down, or burned. For the EU ETS, forestry projects were excluded as potential offsets partly because of the measurement problems and partly because the potential supply of forestry credits is so great that it was feared that the carbon market might be flooded.

Early agricultural credits in the CDM included the capture and combustion of methane (CH_4) from pig manure. As part of its voluntary offset portfolio, TransAlta invested in projects of this type in Latin America. As these are small scale operations, several projects were bundled together by "carbon aggregators" such as Camco, EcoSecurities, and AgCert. AgCert was a particularly dynamic example of this new kind of global entrepreneur, selling a stream of credits expected from captured methane. In other words, AgCert sold the credits forward for several years into the future. Unfortunately, it became known that the technology for handling the methane was poorly maintained and the projects could not deliver what the company had already sold. AgCert became insolvent.

Like any new market, the carbon market has had—and no doubt will continue to have—a range of teething problems.

Lessons to Consider for Designing the Canadian Carbon Market

The EU ETS, as well as voluntary markets like the CCX, provide an excellent opportunity to learn for Ottawa and the provinces when they eventually sit down together and design a carbon market for Canada

First, not everything has to be done at the beginning. The EU so far is tackling only carbon dioxide and only from stationary emitters. The bulk of the GHG reductions are coming from the electrical utility sector, and this is subject to little international (i.e. uncapped) competition. Canada may not have the luxury of this choice if America moves more aggressively, targeting, for example, all six GHGs and a broader swath of the economy.

When the EU ETS opened for business in January 2005, it was working on an estimated baseline for emissions for the 11,000 installations in the system. Indeed, the third-party verification companies were still in the process of being registered. It was still a scramble to put them in place to complete the emission audits for the first true-up date in March 2006. One reason for the delay was that the verifiers had to be registered *separately* in each member state of the EU, and even separately by province in some countries like Spain. It is to be hoped that Canada will not have to follow a similar procedure at the provincial level. Fortunately Ottawa has required large stationary emitters to complete their GHG baseline for 2006 by 2008. So the baseline should be less of a problem in Canada.

Regarding the timeline for targets, expectations have evolved since the three- and four-year targets for the EU ETS. It is widely recognized that these time periods do not mesh with the investment cycle and that longer timelines are more appropriate. In the EU, for example, "the European Commission got member states to agree to its '20-20-20' plan to reduce emissions to 20 percent below 1990 levels by 2020" (*The Economist*, 25 April 2009).

The issue of auctioning some or all of the allowances has also progressed since the EU first issued free allowances for the period from 2005 to 2007. The EU is now auctioning a portion of new allowances, while RGGI has auctioned 100 percent since its inception. In the bill presented by the Obama administration, it was proposed that all the allowances should be auctioned and factored into the federal budget. (However, this was rejected by Congress.) Governments are much less shy about auctioning than they have been in the past. Ottawa should plan to auction most—if not all—of its GHG emission allowances.

The volatility of the price of carbon in the brief history of the EU ETS generated some alarm. For those who fear the impact of a high price of carbon on business, the spikes are alarming. Many favour placing a ceiling on the price by allowing payment into a technology fund in lieu of emission cuts. Both the Alberta legislation and the federal proposals in 2008 advocated such a step. Others fear that the price could fall too low (as happened in Phase One of the EU

ETS), thereby leaving little incentive to cut GHG emissions. This problem could be solved by auctioning all allowances while maintaining a minimum reserve price. Canada's carbon market designers will have to decide whether they want to manage price volatility or not.

One last lesson may be drawn from the first stage of the carbon market. We can surmise that federal inaction on the climate change file in Canada and the US was partly motivated by the fear of what would happen if action were taken that was comparable to the EU ETS. Would it cripple our economy and end the world as we know it? And wouldn't it be pointless if some of the other major uncapped economies did not join the party? This "fear of action" has been largely replaced since the publication of the Stern Report by a "fear of inaction."

The numbers produced by Nicholas Stern's team were overwhelming. The consequences of inaction were much scarier than the costs of taking immediate action to reduce GHGs, whatever the uncertainties in the evolution of the climate and the speed of the human response to the challenge. The worst that critics could throw at the report were quibbles about the discount rates that had been applied to the various cost-benefit scenarios. So, this is a time for even the most timid of decision-makers to creep out of the shadows and prepare for action on the climate change front. In Canada we have wasted twenty years; we cannot afford to waste another twenty.

What Do "Revenue Neutral" and "Carbon Neutral" Really Mean?

During the Liberals' ill-fated 2008 federal election campaign, Stéphane Dion made the "green shift" a central part of the campaign. The heart of the proposal was a carbon tax that would target carbon-intensive activities. (For the probable merits of this concept, see chapter five.) Mr. Dion assured the voters that the tax would be "revenue neutral": the idea here was that the receipts from the tax would all be returned to the voters in various green technology schemes, such as home insulation and so on. Fair enough.

For individual voters, households, and businesses, however, it could not possibly be neutral. For one, the accounting procedures needed to track the impacts of the tax on individuals and households do not exist. For two, the purpose of the carbon tax was to shift behaviour to less carbon-intensive activities. However, until that shift was in place, the tax would be anything but neutral. It was designed to punish carbon-intensive activities like driving an SUV and taking holidays abroad. Otherwise, what would be the point?

In a similar vein, in 2008 the Conservative government promised that any climate change plan would treat all provinces (read Alberta and Saskatchewan) equally, so that no one province would bear a "disproportionate share" of the carbon burden. In Canada's fiscal history, equalization payments have played an important role. Perhaps it would be possible to devise a "carbon transfer" payment system that could help to wean the carbon-intensive provinces off their fossil fuels. In practice, it was difficult to see how this could be done. In a well-structured carbon market, firms will be left to devise their own strategies towards a lower-carbon future. It is very hard to see how this could be done at the provincial level. As Doug Macdonald noted, this type of decision-making "raises all of the regional and continental aspirations and conflicts that have plagued the Canadian federation since 1867."

"Carbon neutral" is another beguiling term that has come into heavy usage in our carbon-constrained world. What does this concept really mean?

The term "carbon neutral" was first used in the corporate context to signify that a corporation had established policies to ensure that in its own operations it would purchase carbon credits to compensate for any GHG emissions that it could not eliminate. There would be a process of verification by an independent third party. This was the state of play until two or three years ago when a more critical approach became widely accepted.

The inventory methodology of the World Resources Institute's GHG Protocol makes distinctions between various "scopes" of emissions. Scope One refers to emissions made directly by a company's processing operations. Scope Two refers to emissions associated with the electricity purchased to run the company's operations. Scope Three involves all other emissions that can be associated with the company—those involved in the supply chain, in the distribution and use of the company's products, in the disposal of the company's products, and—ultimately—in the decommissioning of the company's plant.

In 1998, when BP made its own operational carbon footprint publicly available, it was asked about its wider footprint. What are the GHG emissions associated with the burning of its petroleum products (being about nine times the processing emissions)? The emissions associated with its supply chain? As we have seen, at first BP claimed that the supply chain was beyond its computational control—thousands of firms would have been included. Since then, however, a number of large corporations (headed by Walmart) have undertaken surveys of their supply chains within the framework of the Carbon Disclosure Project. In ten years, the notion of a company's responsibility for its carbon footprint—and hence the notion of carbon neutrality—have broadened considerably.

All of which leads us to an interesting question. When will Canada, as a country, become "carbon neutral"?

The Latest Developments in the Carbon Markets

We know that Canada will have to move quickly, given that the US is already in the process of designing its carbon market. Some politicians in the US will be more than willing to embarrass Canada, if only to reassure their own nervous voters there will be no "free-riding" from countries without caps, like China or Canada. It is an inevitable part of the rhetoric. Consider all the other commodity squabbles we have had between our countries; we can be absolutely certain that a really major issue like climate change will play very big on the North American political scene.

However, it is not only the US that will force the issue. The global carbon market is about to undergo its next major transformation. This is signalled in the following low-key quote from officers of the EU Environmental Commission:

> The CDM should be limited to the poorer developing countries which unfortunately attract a negligible number of investors.... We would envisage that the economically more advanced developing countries would establish domestic cap-and-trade systems by 2020. (Jos Delbeke and Peter Zapfel, An evolving role for international offsets, *Environmental Finance* 10, March 2009)

This is quite a radical statement, especially coming from the diplomacy-laden atmosphere of the EU Environmental Commission. Two things are being proposed here. First, the advanced industrial economies among the "developing countries" are expected to join the GHG-capped countries by the end of the next decade. Second, the CDM will be phased out of these richer developing countries (India, China, Brazil, Indonesia, Mexico, etc.) and restricted to the poorest countries—much of sub-Saharan Africa, Paraguay, Bolivia, Peru, Nepal, southwest Asia, and so on.

This statement, coming from a conservative diplomatic source, must carry weight—and an implicit message for Canada. On our current trajectory (which, as we have noted, is largely going nowhere at the federal level) our climate change efforts could soon be eclipsed by Brazil, Mexico, and other countries which are very much poorer than Canada in *per capita* terms. The clock is not ticking. It is booming.

Chapter 11

The Risks Attendant on Further Delay

The federal government is planning sweeping new climate-change regulations for Canada's electricity sector that will phase out traditional coal-fired power. Any new coal plants will have to include highly expensive—and unproven—technology to capture greenhouse gas emissions and inject it underground for permanent storage, Environment Minister Jim Prentice said in an interview yesterday.

Ottawa also plans to impose absolute emission caps on utilities' existing coal-fired power plants and establish a market-based system to allow them to buy credits to meet those targets, Mr. Prentice said. Electricity users in Alberta, Saskatchewan, and Nova Scotia would be hit hard by the new rules, as their provinces rely on coal for more than 70 percent of their power.

Ottawa takes aim at coal power: Regulations released later this year. *The Globe and Mail*, Report on Business, 29 April 2009

The long-standing "triple E" equation at the heart of our industry is: economic growth = energy growth = emissions. Simply put, this equation must be broken.

TransAlta Corporation, *2007 Report on Sustainability*

Climate risk management ... has to become a normal part of regular activities, in the Bank's client countries, in the Bank's plans, and in its operations.

Ian Burton and Maarten van Aalst, *Look Before You Leap*, World Bank, 2004

Risk Management

When it comes to climate change, we can envision Canada's future in terms of risk management, in the same way that many corporate decision-makers do. The simplest way to categorize the risk to Canada as a country is physical, economic, and reputational.

The physical risks are perhaps the best known. Clearly there are uncertainties. We do not know exactly how to use climate projections to plan infrastructure, manage traffic, or prepare for new threats to public health from heat stress, unknown disease vectors, and contaminated water. Unfortunately, there are still many people in Canada (as elsewhere) who seem to be stuck in the mindset that assumes that decisions that relate to the environment and the economy are structured as a zero-sum game, where winners must be balanced by losers.

A perfect example of this outdated mindset is provided by the province of Saskatchewan:

> One of the few provinces in the black can't afford to go green.
>
> Citing the world economic downturn, Saskatchewan's government is reneging on an election promise to reduce greenhouse gases by 32 percent by 2020.
>
> Environment Minister Nancy Heppner said this week that a 32 percent reduction would be "a pretty huge burden on industry" and that a revised, intensity-based target was forthcoming. (Saskatchewan reneges on climate change, *The Globe and Mail*, 23 April 2009)

In April, the federal government, which committed last year to GHG intensity targets, now embraces absolute reductions. In the same month, Saskatchewan, which also committed last year to absolute reductions, now opts for intensity targets. Confusion reigns in Canadian climate change policy.

What is equally difficult to understand is *why* Saskatchewan is now moving towards "intensity-based" targets in the face of a recession. We saw earlier that in a financial downturn, the demand for power falls and consequently absolute emissions also fall. But intensity-target reductions still have to be met according to the timetable set out in the regulation.

After the publication of the *Stern Review*, there should have been little doubt about the costs of further delays. Yet the spirit of environmental nihilism still hangs heavy. Recall the Canadian myth from chapter one, that whatever we do in Canada won't make any difference to climate change.

There are countless examples of environmental nihilism in action. It is a genuine phenomenon, which increases the difficulty of mobilizing the population. A clear example is the persistently negative attitude towards renewable energy: "it's only 2 percent of the energy mix, so it's too small to make any difference." And yet, recall how British ironmasters at first rejected coal as an alternative fuel to charcoal, circa 1820. (They probably said "it's only 2 percent of the mix" as well.)

Wind power is the favourite target of the environmental nihilist because not only is it intermittent (like solar power) but also wind turbines kill birds. That is quite true. Countless studies have been made of bird deaths attributable to wind turbines, beginning with a famously ill-sited California wind farm that was responsible for the deaths of many American bald eagles in the 1970s. These days, properly sited wind farms are not placed in the pathway of migrating birds, and the deaths from a modern wind farm have been greatly reduced to an average of two birds per turbine per year. Some may feel that number is still too high, but consider that a *great many more* birds are killed by factors such as traffic and the lights of tall buildings. Some concerned groups have mounted campaigns to reduce bird deaths by turning off unused building lights at night. However, the number one killer of birds is the domestic cat.

The intermittency "problem"—like the bird problem—has been studied extensively. Obviously the "smarter" the grid and the better the storage options, the easier it is to manage the fluctuating nature of electricity from wind turbines that feed into the grid. However, we saw on page 48 how even with existing technology, a national or regional grid can produce reliable power with renewables making up 30 percent of the energy mix. This is already happening in Spain and Greece.

(See the UK Energy Research Council's 2006 report, which reviews more than 200 reports and studies worldwide.)

Still, the nihilists will probably stay with the belief that "it's either the economy or the environment" whatever reports are written to the contrary. There are risks to both the environment and the economy, and this chapter considers the likely evolution of these risks, as well as the issue of reputational risk.

Environmental Risks

Some of the environmental risk is domestic, and some is international. Climate change is almost certain to augment the worldwide flows of environmental refugees from vulnerable regions. Those most immediately threatened are the inhabitants of small island states like the Maldives and the island nations of the South Pacific. Australia and New Zealand are already making provisions for increased refugees from these sources.

The largest population mass at acute risk is Bangladesh. The majority of the country's 142 million people is already vulnerable to severe flooding. The regions at risk include areas downstream from the Ganges and the Brahmaputra (due to increased snowmelt in the Himalayas) and from coastal floods driven by cyclones, which may intensify, become more frequent, and affect a wider area under climate change. Many other countries are already vulnerable to increased risk of drought, including Kenya and Morocco. Morocco's rivers are all sourced in snowfields in the Atlas Mountains in the Sahara Desert.

These vulnerabilities will pose policy issues for all wealthy countries, Canada included, in the areas of immigration, international aid, and international security. For over a decade, the CIA—emphatically *not* an environmental NGO—has been citing climate change as a major issue for international security.

Chapter three considered the physical risks to Canada. We can expect further warming in the Arctic, leading to increased stress for the lives and livelihoods of northern peoples. It will also endanger elements of the modern economy, such as pipelines, foundations of buildings, and ice roads for trucks, and increase the risk of floods from ice-dammed rivers.

After the Arctic, the Prairies and the Okanagan, which draw on glacier-fed streams, are particularly vulnerable to climate change. All of Canada—even the warmer regions—enjoys a huge benefit from the free storage of water in the ice and snowfields during the winter. When this water melts, it replenishes aquifers and feeds streams. It is impossible to put a value on this huge hydrological asset as long as we do not pay a realistic price for water. At most, some urban dwellers pay for some of the cost of water delivery and treatment.

We risk losing this asset—which currently renews itself—if we do not pay for water. Just as we need to internalize the cost of GHGs, we also need to internalize the cost of the hydrological cycle as an essential, though currently grossly undervalued, asset.

The complexity of the hydrological cycle in Canada—specifically its responsiveness to a steady increase in temperature in most parts of the country—makes it very difficult to predict the environmental risks inherent in climate change.

Economic Risks

Economic risks may be divided among those sectors of the economy that emit significant quantities of GHGs (and therefore are subject to regulation) and those that are physically most exposed to the impacts of climate change. Some sectors, like oil and gas, fall into both categories. We have seen how the major emitters continue to face a huge amount of uncertainty in light of the ongoing lack of convergence between federal and provincial policy on climate change. Canada's dependence on coal (especially in Alberta, Saskatchewan, and Nova Scotia) in addition to the "on-again, off-again" development of the carbon-intensive oil sands are major issues that have yet to be factored into Canada's climate policy. (We have seen however that this may soon begin to change in response to policy initiatives in the United States.)

In the meantime, federal policy seems to be based on the assumption that a combination of clean coal technologies and carbon capture and storage for the electrical utilities will provide a solution in time to conform with emerging American policy on the carbon content of imports. With oil at $50 to $60 a barrel (as of May 2009), prospects do not look promising for the oil sands.

For those sectors of the economy that are physically vulnerable to climate change, waiting for policies is a dangerous luxury. The effects are already being felt by the forestry sector through the spread of the pine beetle and the impact of forest fires. Agriculture is under threat from water shortages. There may be some relief found in long-overdue applications of irrigation technology and full-cost pricing with improved efficiency. However, as we saw from considering the case of the Okanagan Valley (where irrigation water is still being measured by the acre foot) this is not likely to happen soon. Several other sectors of the Canadian economy, such as health, tourism, electricity transmission, and transportation, are also highly vulnerable.

Reputational Risk

Corporations have begun to investigate their potential for reputational risk under climate change. The Carbon Disclosure Project has highlighted this issue, at least for the larger, publicly listed businesses. In the last decade, expectations for corporate response on the issue have broadened, from considering only a company's internal operations (such as Scope 1 GHG emissions) to computing the impacts of the company's supply chain and the use and disposal of the company's products (Scopes 2 and 3). We are now moving into an era where senior governments in Canada will have to seriously evaluate their reputational risk under climate change.

As long as the EU ETS promised to be the only carbon compliance market in the world, Canada could hide among the rest of the countries without a GHG reduction program. That has now changed. Hurricane Katrina and the *Stern*

Review accomplished what Al Gore's *An Inconvenient Truth* and the IPCC had not. (Everyone has their own choice of tipping factors.) Katrina and Stern appear to have created an unstoppable momentum towards addressing the threat of climate change at a global level. In November 2007 Kevin Rudd led the Australian Liberal Party to victory, partly on a climate change election platform and after eight years of the worst drought in memory. Within a few days Australia had returned to the Kyoto Protocol with a commitment to meet its original reduction target (admittedly a soft one). Over the course of the 2008 US presidential election, it became obvious that the next president would bring the country back into the Kyoto fold too.

Now that these players have returned to the team, others like Canada, Japan, and New Zealand will come under greater scrutiny. As Karan Capoor and Philippe Ambrosi of the World Bank's Climate Change Team diplomatically observed, "although Canada is still officially a party to the Kyoto Protocol, it will follow an alternative reductions schedule (postponing reducing its Kyoto target to beyond 2020)" (Capoor and Ambrosi 2008).

In chapter ten we saw how another significant geopolitical shift is taking place. The bigger and wealthier economies among the developing countries will be expected to articulate reduction targets very soon, perhaps to take effect as early as 2020. If Brazil, Chile, China, India, Mexico, and others announced forthcoming goals, then Canada's position (along with Japan and New Zealand) would become very exposed, slipping from the disappointing to the ludicrous.

These issues are fast approaching. They will be front and centre at the Copenhagen Conference of the Parties to the Kyoto Protocol in December 2009. Unless a revolution occurs in Canadian climate policy between now and then, we will almost certainly place last among the industrialized countries when it comes to climate change preparedness. The best we could hope for would be to share last place with Japan and New Zealand.

This might not be the lowest point for Canada and climate change. If Canada still has no traction on the issue when the richer developing countries become involved in making commitments to global climate change policy then we can expect to be pilloried. If Brazil implements a credible GHG plan and Canada still lacks one, what excuse could we offer? That we are a large, cold country with a complex political structure? Brazil is a large country as well, with a population of 186 million people and a political structure every bit as complex as Canada's. Most tellingly, it has a GDP per capita of US $4,270 compared with Canada's US $34,480. This is reputational risk on a very large scale.

Six months is a short period of time in which to make amends. Fortunately—at least for the optimists among us—the story could still have a happy ending. The final chapter will consider the opportunities presented by climate change. Perhaps carrots will be more effective than sticks.

Chapter 12

Opportunities in Climate Change

In October, Enbridge Gas Distribution Inc. opened a 2.2 megawatt hybrid fuel-cell power generation station in Toronto, an efficient low-emissions plant that is the first of its kind in the world.

Our aging population, climate change, and emerging technologies spell opportunity. *The Globe and Mail*, Report on Business, 29 April 2009

"I'm optimistic if we're sensible," Lord Stern said. "We can move pretty quickly to control emissions, to develop the new technologies, to put in the economic policies that are going to hold the costs down. And I do think it would help pull us out of a recession."

A green offshoot of recession. *The Globe and Mail*, Report on Business, 1 May 2009

I would be concerned if we put all our emphasis on the macro level that has consumed policy dialogue to date. We do need a macro framework but we must very strongly supplement it with precision policies that directly require the use of renewable energy, require improvements in energy efficiency, and increase performance standards for consumer products and automobiles.

Those policies are crucial if we are going to see the rapid infusion of new technologies into our economy and reductions in greenhouse gases.

Kathleen McGinty in an interview with Christopher Cundy, *Environmental Finance*, March 2009, 36

Climate Change as a Crisis

It is no exaggeration to describe climate change as a "crisis." This is not hyperbole: crisis is defined by the OED as "a time of severe difficulty or danger." This is exactly what climate change represents for the human population, as well as many mammals, birds, and trees. Fish populations might do much better if their human predators disappeared from the planet. The bacteria certainly wouldn't notice our passing.

The climate change crisis has implications on a scale comparable to a conventional world war or a plague like the fourteenth-century Black Death, estimated to have killed one-third of the population of Western Europe. The effects would be less sudden, but they would last considerably longer. If we ceased emitting GHGs tomorrow, the warming would continue for at least a hundred years, and perhaps longer, depending on what changes happen to the permafrost, forests, and so on.

We have been reading about "our overloaded planet" for several decades now. With a population of nearly 6.8 billion, there is little room to manoeuvre. Globally we produce enough food to feed everyone; but poor distribution means that many people live on the brink. A major disturbance like a shift in rainfall pattern could have very large consequences for human morbidity and mortality.

"Crisis" comes from the Greek word "krisis," meaning decision. Much of this book has been about making decisions (or in the Canadian context, *not* making decisions). Perhaps this does not sound like a time of great opportunity.

Why Opportunities?

At the simplest level, the crisis will create demand for new products. Examples that spring to mind include financial products and renewable energy. We are facing new problems, and we will need new solutions. That may sound simplistic—but need has always been a driver for innovation.

Carbon reduction credits, weather derivatives, and catastrophe bonds are examples of new financial products that have been developed in response to climate change. If you had proposed selling these products fifteen years ago, nobody would have known what you were talking about. Today, they are well established in the financial markets.

This book has made frequent reference to the "carbon market" that trades GHG emission reduction credits both in the allowance (or compliance) market in the European Union Emissions Trading Scheme (EU ETS), as well as through the Clean Development Mechanism and Joint Implementation. In 2008 the carbon market traded US $118 billion in value.

Weather derivatives are based on expected weather parameters, such as Heating Degree Days and Cooling Degree Days; these are designed to allow energy providers to hedge their exposure to adverse or unusual weather. A company selling electricity to run air conditioners would consider "adverse weather" to be a cool summer with little demand for cooling. The energy provider can buy a derivative that offers a payout if the summer (at a given location) is cooler than expected. The "weather market" is also a multi-billion dollar market and covers a variety of adverse conditions including snow, frost, wind, and drought.

Catastrophe bonds—also known as "cat bonds"—were developed in the wake of Hurricane Andrew (1992) to transfer some of the insurers' exposure to extreme weather events such as hurricanes. Insurers worldwide were left seriously exposed after the payout for Hurricane Andrew, for which there was no precedent or expectation in the insurance industry. A partial solution was for insurers to sell catastrophe bonds to the capital markets: these transferred some of the risk to investors, where the higher rate of return justified the greater risk. The first cat bond was launched in 1994. None were triggered by an extreme event until Hurricane Katrina in 2005. For a decade the investors who took on the new risk made an enhanced rate of return, while the insurers hedged their exposure to extreme events.

Opportunities are presented by the climate change crisis because new products and new ways of thinking are needed. For example, we now have energy companies whose "product" is to reduce your energy needs, not sell you more energy. For Western material society, the idea that "less is better" is a revolutionary way of thinking.

Technological Opportunities

The transition to a low-carbon economy will be as far-reaching as the original switch to fossil fuels. It must also take place much more quickly if we are to avert "dangerous" climate change. It is unlikely that we will discover a single breakthrough technology that will make this transition swift, painless, and profitable. Therefore, placing all our expectations on one fortuitous discovery is clearly unwise. Much of the talk about "carbon capture and storage" and "clean coal" leans too far in this direction, a mistaken belief which is sometimes known as "technological salvation." Fifteen years ago, the fuel cell was hailed as just such a miracle, a technology that we could swap for our internal combustion engines and drive on as before. So far, the impact of the fuel cell on road transportation has been much more modest.

To make the transition in good time, governments in Canada and elsewhere must help reduce the uncertainty for innovators. In Canada, most governments have done just the opposite. There is enough uncertainty in the marketplace already without needlessly creating more. Recall, for example, recent fluctuations in the price of oil and the recession.

The general requirements for moving to a low-carbon economy are clear. We need to reduce demand for activities that require energy. And what we do use must be managed more efficiently. We must switch from fossil fuels to renewables. We can continue to debate the place of nuclear power and large-scale hydroelectric power in the energy mix.

The important first step in reducing demand for energy is to establish a price that includes the cost of emitting greenhouse gases. The EU ETS began this process in 2005, following various voluntary bodies like the Chicago Climate Exchange in 2003. The price signal may be supported by a number of measures that include tax breaks for development costs, loans from a publicly supported technology fund, and renewable energy requirements for all energy producers. For renewable energy a higher feed-in tariff has been critically important in those jurisdictions (mostly in the EU) where renewables have quickly become a significant part of the energy mix.

In most industrial economies there has been a grave reluctance to see the prices of energy and water rise; this was true well before the spectre of climate change. Both are considered so fundamental to our well-being that they continue to be widely subsidized. Only the twin fears of climate change and supply insecurity have brought these long-standing policies into question.

Public intervention in technological change is needed to accelerate the low-carbon transition, particularly in urban and transportation planning. There is no more obvious waste of energy than sitting in a traffic jam—a daily occurrence in most North American cities. It has been calculated that only half the energy used in urban road transportation is used for moving forward; the rest is wasted while idling. Congestion charges are a means of reducing this wasted time and energy while providing funding to improve public transport. Another efficiency gain will come from charging more for peak afternoon/evening electricity, when it is more expensive to generate. Time-of-day meters are gradually being introduced.

Water supply and treatment also uses considerable energy, sometimes accounting for more than half the municipal energy budget. Many municipalities in Canada do not yet use volumetric pricing, while those that do so are still far from full-cost recovery. It is likely that the impacts of climate change will accelerate improvements in the management of energy and water. The sooner these changes are made, the better.

The technological transformation needed in a low-carbon economy involves many interrelated steps. Some of these will require innovations such as improved energy storage and the development of rapid recharging systems for vehicles. However for many of these steps, the technology already exists. Moreover, many of these changes confer multiple advantages in addition to addressing climate change. Reducing or eliminating traffic congestion brings immediate improvements of air quality, and quality of life generally.

Economic Opportunities

The link between the challenges posed by climate change and the recession was made early on. We have heard a great deal about how investment in green infrastructure can help us escape the recession, just as investment in public infrastructure created jobs during the Great Depression.

Nicholas Stern pointed out that "we can move pretty quickly to control emissions, to develop the new technologies, to put in the economic policies that are going to hold the costs down. And I do think it would help pull us out of a recession." Some critics caution that increasing government involvement is a step backwards, to the nationalization of industries following World War II. Many fear that governments tend to make poor decisions when they interfere in the marketplace; recently cited examples include subsidies for corn-based ethanol products (held responsible for food-price increases as well as producing GHG emissions equivalent to those of petroleum).

Despite differences of opinion, there is widespread hope that the low-carbon future will create a range of "green-collar" jobs throughout the economy—agriculture, forestry, utilities, transportation, manufacturing, construction, health, education, finance, information technology, as well as new jobs in the carbon market itself. There is certainly reason to hope that, although the transition may cost us some jobs as energy, water, and materials increase in price, new jobs will also be created. The tragedy for Canada is that all the time that we have delayed taking action, we have been missing out on all these innovative opportunities. Until we put a price on carbon we will not be able to benefit from participation in the developing low-carbon economy. Even worse, we may be driven out of established markets, most significantly the US, if we continue on a high-carbon trajectory.

Political Opportunities

To appreciate the political opportunities that climate change presents, let us recall the distinguishing characteristic of GHGs in the atmosphere. We noted earlier the long "residence time" of these gases: methane remains in the atmosphere for sixty years, carbon dioxide for a hundred, and the other gases for even longer. That's why this is a long-term problem that we are passing on to our children, our grandchildren, and even their children. While the GHGs are in the atmosphere all those decades, they become well mixed and therefore an inextricably global problem. How can this be a political opportunity?

The opportunity lies in the fact that this global problem has consequences for every country in the world, and for every region of every country. Some countries are more vulnerable, and may see the effects sooner, than others—but no single country is immune. As this reality becomes more apparent, all countries will have to come to the bargaining table in good faith. Climate change is not a problem like poverty, world trade, or pollution. It is not somewhere "over there"; nor does it lie in the distant future.

It may seem idealistic to assume that because this is a truly difficult, global problem, world leaders and citizens will somehow converge to work out a cooperative solution. Yet we have seen a precedent: recall how urban mortality rates during the Industrial Revolution exceeded birth rates (page 25). This imbalance was not reversed until philanthropists, politicians, and medical professionals realized that the underlying problem was the poverty and poor hygiene in which most urban dwellers lived. Infectious diseases were endemic, and endangered rich and poor alike. The answer, "The Public Health Idea," came with the understanding that there was no solution for the rich *unless there was also a solution for the poor*. (Steven Johnson's *The Ghost Map* [2006] recounts this fascinating story.)

That was a problem on the urban scale. Climate change operates on a global scale. A real solution must involve tangible and immediate benefits to the poorer countries of the world.

There is no "gated community" solution to protect rich people while ignoring the poor.

The subtitle of Nicholas Stern's book *The Global Deal* is optimistic: *Climate Change and the Creation of a New Era of Progress and Prosperity*. He also concludes on an optimistic note, which I hope will inspire policy makers in Canada to move ahead rapidly to play their part and face up to the challenge of climate change:

> We know what we have to do; the prize is enormous. The citizens and politicians of the world, community by community, nation by nation, will now determine whether we can create and sustain the international vision, commitment, and collaboration which will allow us to take this special opportunity and rise to the challenge of a planet in peril. (Stern 2009)

Glossary

Terms that are bolded in the text are listed elsewhere in the glossary. "Greenhouse gas" has been abbreviated "GHG."

Absolute emission limits
Caps on GHGs that are stipulated per period of time and reduced in each subsequent period.
Adaptation
Changes made by a society to respond to the impacts of climate change.
Allowance market (for carbon credits)
This term is used to describe restrictions placed on industry and the market that is created for reduction credits. The **cap-and-trade system** is a prime example of this.
Antecedent soil conditions
The level of water saturation in the ground, a factor that in combination with heavy rainfall can contribute to flooding.
Auctioning
Carbon allowances may be auctioned to the highest bidder (as in the Regional Greenhouse Gas Initiative), as opposed to being distributed free-of-charge (as in Phase One of the European Union Emissions Trading Scheme).

Banked allowances
Carbon allowances may be kept, or banked, from one year to the next. Longer banking periods will reduce price volatility.
Baseline (for GHG reductions)
Reductions are made with reference to emission levels from a baseline year. For the **Kyoto Protocol** the baseline year is 1990.

Biodiversity
The variety of species in a given area, e.g. tree species per hectare.
Business-as-usual
The baseline scenario for Global Circulation Models, determined by the assumption that no **mitigative** action is taken to reduce GHGs.

Cap-and-trade system
A carbon market in which GHG emissions are capped and surplus credits may be traded among the capped installations.
Carbon aggregator
A company that sources carbon credits from GHG reduction projects in uncapped countries and bundles them into larger units for investors or capped installations.
Carbon burden
Many activities generate GHG emissions, but the burden (or obligation) for reductions is usually placed on only some of these activities.
Carbon capture and storage
For some GHG-emitting activities, technology is being developed to remove (or capture) the gases before they are emitted into the atmosphere and then store them underground. Prototypes exist for this type of process for electrical utilities.
Carbon-constrained world
This is a world in which carbon carries a price.
Carbon credits or allowances
At the inauguration of a carbon market the regulator distributes credits, or allowances, to the capped installations.
Carbon cycle
The passage of all forms of carbon from sources (such as the combustion of fossil fuels) to **sinks** (such as the atmosphere, the ocean, biomass, and soils).
Carbon economy
Any economic activity that is connected to activities dedicated to the reduction of GHGs.
Carbon finance
The use of market instruments to reduce GHGs and transfer risks associated with the **carbon economy**.
Carbon footprint
A total set of the GHGs associated with a product, individual, or organization, representing all emissions created directly or indirectly.
Carbon funds
Funds that are invested in GHG reduction activities.
Carbon neutral
The attainment of a status whereby all GHG-emitting activities are offset by investments in GHG reduction.
Carbon offset credits
GHG emissions may be balanced (or offset) by GHG-reducing activities

such as afforestation or investment in renewable energy. The reductions may be accorded credits by a regulatory body.

Carbon sequestration

The two main ways to sequester (or bury) carbon are to plant trees or to inject the carbon underground.

Carbon tax

A tax on goods and activities according to the quantity of GHGs they embody or emit.

Carbon trading

The trading of GHG reduction credits and associated offsets.

Catastrophe bonds

Investment bonds that are issued against the occurrence of a natural catastrophe (or set of catastrophes), such as an earthquake or a hurricane. If the catastrophe occurs then the investors pay their investment income stream, and sometimes the principal sum, to the issuer of the bond.

Cleantech

Short form for "clean technology," such as an energy- or water-efficient process.

Climate change commitment

Changes in climate to which the world is committed, due to the GHGs that have already been emitted.

CO_2 equivalent

Under the **Kyoto Protocol** six GHGs, including CO_2, are targeted for reduction. Because of their different chemical properties, the gases vary in their capacity for trapping heat in the atmosphere. Calculations have been made to estimate the "global warming potential" of each gas over a hundred-year period in order to establish the impact of reducing each gas compared with reducing CO_2.

Combined heat and power

Power stations that utilize the heat energy produced during the generation of electricity. In standard power stations this heat is normally vented to the atmosphere.

Compliance market (for carbon credits)

The trading of carbon credits that have been created in order to comply with GHG reduction regulations. The largest compliance market is the European Union Emissions Trading Scheme.

Congestion charges

Charges that the motorist must pay in order to enter a central zone of a city. The money collected is usually directed towards the improvement of public transport. The largest example of this is in London; the oldest is in Singapore.

Contraction and convergence

A proposal to reduce global GHG emissions in which every country converges on the same per capita allowance for emissions. The rich countries would reduce their per capita emissions, while poorer countries could increase them.

Counterparty risk
Like all trading, **carbon trading** is exposed to the risk of the counter party (buyer or seller) failing to complete its side of the trade. As the **carbon economy** is in a fledgling state, counterparty risk is high.

Demand elasticity
The impact of a change in price on the level of demand for goods and services. Where demand is elastic, responsiveness is high.
Demographic transition
The transition in human populations from high birth rates and high death rates to low birth rates and low death rates. This usually occurs as a society becomes wealthier.
Differentiated targets for GHG reduction
The recognition that poorer countries should have lower GHG reduction targets than richer countries.

Earth Summit
The conference that took place in Rio de Janeiro in 1992. It established the United Nations Framework Convention on Climate Change, which in turn led to the **Kyoto Protocol**.
Emissions trading
The trading of emission reduction credits, beginning with the US acid rain reduction program in the 1990s.
Environmental finance
The use of market instruments to achieve environmental objectives and transfer environmental risk.
Environmental nihilism
The assumption that efforts to achieve environmental goals—such as stabilizing the climate—are doomed to failure.
Environmental refugee
Refugees who are forced from their homes and countries by adverse environmental conditions, such as drought or hurricanes.
Evapotranspiration
The transfer of water vapour from the ground and from living organisms to the atmosphere in response to warm temperatures. This term combines evaporation from the ground with transpiration through plants.
Exchange trading
The trading of investments through an exchange, like the Toronto Stock Exchange. It is an indicator that a market has matured as it has reduced its reliance on **over-the-counter trading**. Exchange trading provides price transparency, increases liquidity in the market, and reduces **counterparty risk**.

Feedbacks, negative and positive
In the language of systems analysis, negative feedbacks reduce the impact

of a process (such as climate change) while positive feedbacks augment the process.

Feed-in tariffs

The establishment in some countries of guaranteed prices for energy that is sold into the grid, to encourage investment in renewable energy.

Flood plain

The area that is expected to be flooded (based on historical data) over a certain time period. For example, data will determine the 20-year flood plain, the 50-year flood plain, and so on.

Food miles

Describes the component of a food product's **carbon footprint** associated with its transportation to market.

Fossil fuels

Fuels based on materials that have decayed over millennia, and hence cannot be renewed. The principal fossil fuels are coal, lignite, petroleum, and natural gas.

Fossil lakes

In the Canadian context this term refers to lakes that were formed at the end of the last ice age. Their annual rate of replenishment in the current climate is very small compared with their volume.

Freeze-thaw cycle

In some regions the winter brings alternate periods of freezing and thawing. This alternation is very destructive to infrastructure, especially road surfaces. Under climate change, the persistently frozen state of many Canadian regions during winter will be replaced by this cyclical temperature swing.

Freezing rain

Freezing rain occurs when a warm air mass is forced aloft on encountering a dense, cold air mass. Rainfall occurs as the warmer air rises, and as this rain falls through the underlying colder air it cools rapidly. The resulting particles then freeze on contact with surfaces such as buildings and hydro towers.

G8

The Group of Eight, a forum of eight large, industrialized nations: Canada, France, Germany, Italy, Japan, Russia, the United Kingdom, and the United States.

General Circulation Model

A computer-based simulation model of atmospheric processes developed to help us understand the possible implications of climate change under various economic and **mitigative** scenarios.

Green buildings

New or retrofitted structures, designed to reduce demand for energy and water.

Greenhouse gases

Water vapour and certain other gaseous components of the atmosphere that trap the heat radiating from the Earth's surface, creating a blanket, or greenhouse, effect to warm the surface. The six greenhouse gases regulated under the Kyoto Protocol are carbon dioxide, methane, nitrous oxide, hydrofluorocarbons, perfluorocarbons, and sulphur hexafluoride.

Heat island effect
Buildings and other surfaces trap the sun's heat during the day and radiate it back at night, increasing the ambient temperature by several degrees. Hard surfaces (buildings, roads) absorb more heat than vegetation and hence radiate more back at night. This differential means that cities are warmer than the surrounding countryside, thus creating a "heat island effect."

Heat response plans
Summer heat in the city has been a health hazard for decades. As cities grow, they produce a more marked **heat island effect.** Global warming associated with climate change enhances the impact of heat stress in the summer. Several Canadian cities have developed heat response plans to attempt to protect the vulnerable members of society from enhanced heat stress.

Ice roads
Seasonal highways in the Canadian Arctic built on compacted snow and ice. Warmer conditions will reduce the seasonal availability of this transport mode.

Intensity emission limits
Caps on GHGs that are stipulated per unit of output, with the quantity of GHGs per unit decreasing over time.

Intermittency issue
The concern that energy from a renewable source may be generated at irregular intervals. However, when renewable energy is fed into a grid the fluctuating inputs can be managed so that the level of power delivered by the grid is constant.

Investment cycle (in industrial plants and equipment)
Every business operates according to a known cycle of investments in new plants and equipment maintenance. GHGs can be reduced at the least cost when the reduction targets mesh with the cycle of new investment.

Kyoto Protocol
A GHG-reduction agreement that was reached in Kyoto, Japan, in 1997. The Protocol obliged industrialized countries to reduce their emissions relative to a 1990 **baseline**. Canada ratified the Protocol in 2002, and it came into force in 2005.

Land-use change
Land-use changes, particularly deforestation, account for approximately 20 percent of GHG emissions worldwide. GHG emissions may be reduced

by reforestation, conversion of arable land to pasture, and the introduction of minimum tillage on arable land.

Low-carbon technology

Technologies that reduce GHG emissions through improved efficiency, fuel switching, or innovative new processes.

Market fundamentals

Supply and demand, the fundamental drivers of price for a product.

Market sentiment

Investors acting in anticipation of a believed change to the **market fundamentals**.

Methane

A GHG (CH_4), sometimes referred to as natural gas.

Methane capture

The reduction of methane emissions by a variety of activities, such as the conversion of landfill leakage to energy, the capture of emissions from agricultural waste, or the capture of methane leaking from coal mines.

Mitigation

A set of activities designed to reduce GHG emissions.

Modal split

The use of various modes of transportation—train, bus, private automobile, cycling, walking. For traffic management and GHG emissions management a key variable is the modal split for daily commuting.

Oil sands

Oil found as bitumen in shale and sand, as in Alberta and Saskatchewan. The oil that is extracted and refined from such sources produces more GHG emissions than conventional oil. This is usually referred to by critics as "dirty oil" or "the tar sands."

Over-the-counter trading

The trading of investments bilaterally between a buyer and a seller. The price and conditions of the trade are confidential.

Payback time

The time taken for an investment to pay for itself in terms of savings in energy and other recurrent costs.

Permafrost

Permanently frozen ground, widespread in the Canadian North. Climate change is melting the permafrost, leading to landslides and the release of **methane**.

Project market (for carbon credits)

This term is used to describe GHG reduction projects in countries that do not have GHG caps. Once approved, the GHG reductions may be available for sale into capped countries.

Renewable energy
Sources of energy that are not reduced by usage, such as wind, solar, tidal, and wave power. Geothermal energy is considered renewable although it may be depleted through over-extraction. "Small" hydro, especially when run-of-the-river, is considered renewable (although definitions of what is "small" may be disputed). In the UK, energy derived from landfill **methane capture** was accepted for Renewable Energy Quotas, although the project lifetime is typically only 30 years.

Residence time (in the atmosphere)
Gases emitted into the atmosphere stay there for a given length of time. For acid rain precursors, this is typically a few days. In contrast, the shortest residence time for a GHG is approximately 60 years (for **methane**).

Sea-level rise
Climate change will produce a rise of sea level through thermal expansion of the ocean and the melting of glaciers, sea ice, and land ice.

Sink
A feature or system that absorbs or contains compounds (such as GHGs). Examples of carbon sinks in the environment include the atmosphere, oceans, biomass, and soils/sediments.

Somatic energy
The bodily energy of humans and animals.

Technological salvation
A belief that a major problem (such as climate change) can be solved by the development of new technology without requiring any significant change in human behaviour.

Transaction costs
The cost of a business transaction, such as selling a carbon credit. In the early stages of the development of a market, transaction costs are high. As a market grows and matures, transaction costs should fall.

Transaction log
A record of all trading done in the carbon market. Each jurisdiction must establish such a log to ensure the validity and ownership of carbon credits.

Venture capital
Capital that is available for start-up companies such as those in the **renewable energy** sector. Investors expect to make a return on their investment from the sale of the company after a few years through an "initial public offering" on the stock exchange or from selling to an established company in a related sector.

Voluntary market (for carbon credits)
Individuals and companies that invest in carbon credits even when not bound to do so, either for altruistic reasons, for a public relations benefit,

or in anticipation of being regulated in the foreseeable future. The largest voluntary market is the Chicago Climate Exchange.

Weather derivatives
Financial instruments based on some aspect of the weather, such as temperature, rainfall, snow, frost, and wind. Investors buy weather derivatives in order to hedge their exposure to weather that is unfavourable for their line of business.

Zero-sum game
A situation for which the benefits from an outcome are fixed, so that as much as one party benefits, the other must lose. Pessimists about climate change (see **environmental nihilism**) believe that any benefit from GHG **mitigation** must come at a cost commensurate with the benefit.

Further Reading

Chapter 1

Flannery, T. 2005. *The Weather Makers: How We Are Changing the Climate and What It Means for Life on Earth*. Toronto: HarperCollins.

Weaver, A. 2008. *Keeping Our Cool: Canada in a Warming World*. Toronto: Viking Canada.

Chapter 2

Environment Canada. 1988. Conference Statement, "The Changing Atmosphere: Implications for Global Security." Toronto.

Firor, J. 1990. *The Changing Atmosphere: A Global Challenge*. New Haven: Yale University Press.

Ponting, C. 1991. *A Green History of the World*. Harmondsworth: Penguin Books

Vanderburg, W.H. 2000. *The Labyrinth of Technology: A Preventative Technology and Economic Strategy as a Way Out*. Toronto: University of Toronto Press.

Chapter 3

Natural Resources Canada. 2008. *From Impacts to Adaptation: Canada in a Changing Climate 2007*. Available from http://adaptation.nrcan.gc.ca.

Stern, N. 2009. *The Global Deal: Climate Change and the Creation of a New Era of Progress and Prosperity*. New York: Public Affairs.

Chapter 4

Capoor, K., and P. Ambrosi. 2008. *State and Trends of the Carbon Market, Executive Summary.* World Bank. Entire report available from http://www.worldbank.org/reference.

Fenhann, J. 2007. "Clean Development Mechanism: Counting the CERs." *Environmental Finance* 9 (November).

Grubb, M., C. Vrolijk, and D. Brack. 1999. *The Kyoto Protocol: A Guide and Assessment*. London: Earthscan.

Chapter 5

Boyle, G., B. Everett, and J. Ramage. 2003. *Energy Systems and Sustainability: Power for a Sustainable Future*. Oxford: Oxford University Press.

Harvey, D. 2006. *A Handbook on Low-Energy Buildings and District-Energy Systems*. London: Earthscan.

UK Energy Research Centre. 2006. *The Costs and Impacts of Intermittency: Key Issues, Main Findings*. A UKERC Technology and Policy Assessment Report. Available from http://www.ukerc.ac.uk.

White, R.R. 2001. *Building the Ecological City*. Cambridge: Woodhead Publishing.

Chapter 6

Canadian Council of Ministers of the Environment. 2003. *Climate, Nature, People: Indicators of Canada's Changing Climate*. Ottawa: CCME.

Cohen, S., D. Neilsen, and R. Welbourn, eds. 2004. *Expanding the Dialogue on Climate Change and Water Management in the Okanagan Basin, British Columbia*. Environment Canada, Agriculture and Agri-Food Canada, and the University of British Columbia.

Keating, M. 1997. *Canada and the State of the Planet: The Social, Economic and Environmental Trends that are Shaping Our Lives*. Toronto: Oxford University Press.

Natural Resources Canada. 2008. *From Impacts to Adaptation: Canada in a Changing Climate 2007*. Available from http://adaptation.nrcan.gc.ca/.

Chapter 7

Government of Canada. 2006a. *Canada's 2006 Greenhouse Gas Inventory: A Summary of Trends*. Available at http://www.ec.gc.ca/pdb/ghg/inventory_report/2006/som-sum_eng.cfm.

Government of Canada. 2006b. *Canada's Fourth National Report on Climate Change: Actions to Meet Commitments under the United Nations Framework Convention on Climate Change*. Available from http://www.ec.gc.ca/publications.

Government of Canada. 2008. *Turning the Corner: Canada's Plan to Reduce Greenhouse Gas Emissions and Air Pollution*. Available from http://www.ec.gc.ca/publications.

Chapter 8

Canadian Institute of Chartered Accountants. 2008. *Climate Change and Related Disclosures, Executive Briefing*. Available from http://www.cica.ca.

Conference Board of Canada. 2007. *Carbon Disclosure Project Report 2007: Canada 200*. Available from http://www.cdproject.net.

TransAlta Corporation. 2008. *2007 Report on Sustainability*. Available from http://www.transalta.com.

Chapter 9

Capoor, K. and P. Ambrosi. 2008. *State and Trends of the Carbon Market, Executive Summary*. World Bank. Available from http://www.worldbank.org/reference.

Government of Canada. 2006. *Canada's Fourth National Report on Climate Change: Actions to Meet Commitments under the United Nations Framework Convention on Climate Change*. Available from http://www.ec.gc.ca/publications.

Chapter 10

Delbeke, J., and P. Zapfel. 2009. An Evolving Role for International Offsets. *Environmental Finance* 10 (March).

Gonzalez, G. 2009. US Outlines Proposals for National GHG Reporting. *Environmental Finance* 10 (April).

Chapter 11

Burton, I., and M. van Aalst. 2004. "Look Before You Leap: A Risk Management Approach for Incorporating Climate Change into World Bank Operations." *Environment Department Working Paper No. 100*. Washington, DC: World Bank.

Chapter 12

Cundy, C. 2009. Catastrophe Bond Market Bounces Back. *Environmental Finance* 10 (April).

———. 2009b. Weather Risk Volumes Drop as Speculators Rein in Activity. *Environmental Finance* 10 (April).

———. 2009c. Profile: Kathleen McGinty. *Environmental Finánce* 10 (March).

Johnson, S. 2006. *The Ghost Map: The Story of London's Most Terrifying Epidemic and How it Changed Science, Cities and the Modern World*. Harmondsworth: Penguin Riverhead Books.

Stern, N. 2009. *The Global Deal: Climate Change and the Creation of a New Era of Progress and Prosperity*. New York: Public Affairs.

References

Boyd, D.R. 2003. *Unnatural Law: Rethinking Canadian Environmental Law and Policy*. Vancouver: University of British Columbia Press.

Boyle, G., B. Everett, and J. Ramage. 2003. *Energy Systems and Sustainability: Power for a Sustainable Future*. Oxford: Oxford University Press.

Brun, S.E., et al. 1997. *Coping with Natural Hazards in Canada: Scientific, Government and Insurance Industry Perspectives*. A study written for the Round Table on Environmental Risk, Natural Hazards and the Insurance Industry. Environmental Adaptation Research Group, Environment Canada, and the Institute for Environmental Studies, University of Toronto.

Burton, I. 2008. Moving forward on adaptation. In *From Impacts to Adaptation: Canada in a Changing Climate*, ed. D.S. Lemmen et al. Ottawa: Government of Canada.

Burton, I., and M. van Aalst. 2004. Look Before You Leap: A Risk Management Approach for Incorporating Climate Change in World Bank Operations. *Environment Department Working Paper No. 100*. Washington, DC: World Bank.

Canadian Council of Ministers of the Environment. 2003. *Climate, Nature, People: Indicators of Canada's Changing Climate*. Ottawa: CCME.

Canadian Institute of Chartered Accountants. 2008. *Climate Change and Related Disclosures*, Executive Briefing. Available from http://www.cica.ca.

Capoor, K., and P. Ambrosi. 2008. *State and Trends of the Carbon Market*, Executive Summary. World Bank. Available from http://www.worldbank.org/reference.

Cohen, S., D. Neilsen, and R. Welbourn, eds. 2004. *Expanding the Dialogue on Climate Change and Water Management in the Okanagan Basin, British*

Columbia. Environment Canada, Agriculture and Agri-Food Canada, and the University of British Columbia.

Conference Board of Canada. 2007. *Carbon Disclosure Project Report 2007: Canada 200*. Available from http://www.cdproject.net

Cundy, C. 2009a. Catastrophe Bond Market Bounces Back. *Environmental Finance* 10 (April).

———. 2009b. Weather Risk Volumes Drop as Speculators Rein in Activity. *Environmental Finance* 10 (April).

———. 2009c. Profile: Kathleen McGinty. *Environmental Finance* 10 (5).

Delbeke, J. and P. Zapfel. 2009. An Evolving Role for International Offsets. *Environmental Finance* 10 (March).

Environment Canada. 1988. Conference Statement, "The Changing Atmosphere: Implications for Global Security," Toronto.

Farrar, J.L. 1995. *Trees in Canada*. Markham: Fitzhenry & Whiteside.

Fenhann, J. 2007. Clean Development Mechanism: Counting the CERs. *Environmental Finance* 9 (November).

Firor, J. 1990. *The Changing Atmosphere: A Global Challenge*. New Haven: Yale University Press.

Flannery, T. 2005. *The Weather Makers: How We Are Changing the Climate and What It Means for Life on Earth*. Toronto: HarperCollins.

Fogarty, C. 2003. Why Did a Category-Two Hurricane Hit Nova Scotia? An Explanation of the Unusual Intensity of Hurricane Juan. Canadian Hurricane Centre, Environment Canada. Available at http://www.atl.ec.gc.ca/weather/hurricane/juan/intensity_e.html.

Furgal, C., and J. Seguin. 2006. Climate Change, Health and Vulnerability in Canadian Northern Aboriginal Communities. *Environmental Health Perspectives* 114 (12).

Gonzalez, G. 2009. US Outlines Proposals for National GHG Reporting. *Environmental Finance* 10 (April).

Government of Canada. 2006a. *Canada's 2006 Greenhouse Gas Inventory: A Summary of Trends*. Available at http://www.ec.gc.ca/pdb/ghg/inventory_report/2006/som-sum_eng.cfm.

———. 2006b. *Canada's Fourth National Report on Climate Change: Actions to Meet Commitments under the United Nations Framework Convention on Climate Change*. Available from http://www.ec.gc.ca/publications.

———. 2008. *Turning the Corner: Canada's Plan to Reduce Greenhouse Gas Emissions and Air Pollution*. Available from http://www.ec.gc.ca/publications.

Grubb, M., C. Vrolijk, and D. Brack. 1999. *The Kyoto Protocol: A Guide and Assessment*. London: Earthscan.

Harvey, D. 2006. *A Handbook on Low-Energy Buildings and District-Energy Systems*. London: Earthscan.

Intergovernmental Panel on Climate Change. 2007. *Fourth Assessment Report: Climate Change 2007*. Synthesis report available from http://www.ipcc.ch.

Johnson, S. 2006. *The Ghost Map: The Story of London's Most Terrifying Epidemic and How it Changed Science, Cities and the Modern World*. Harmondsworth: Penguin Riverhead Books.

Keating, M. 1997. *Canada and the State of the Planet: The Social, Economic and Environmental Trends that are Shaping Our Lives*. Toronto: Oxford University Press.

Labatt, S., and R.R. White. 2002. *Environmental Finance: A Guide to Environmental Risk Assessment and Financial Products*. Hoboken, NJ: Wiley.

———. 2007. *Carbon Finance: The Financial Implications of Climate Change*. Hoboken, NJ: Wiley.

Macdonald, D. 2007. *Business and Environmental Politics in Canada*. Peterborough, ON: Broadview Press.

Macdonald, D., et al. 2002. *Ratification of the Kyoto Protocol: A Citizen's Guide to the Canadian Climate Change Policy Process*. Toronto: Environmental Studies Program, Innis College, University of Toronto.

Mirza, M.O.Q. 2004. *Climate Change and the Canadian Energy Sector: Report on Vulnerability, Impact and Adaptation*. Ottawa: Environment Canada.

National Round Table on the Environment and the Economy. 1999. *Canada's Options for a Domestic Greenhouse Gas Emissions Trading Program*. Ottawa: NRTEE.

———. 2005. *Economic Instruments for Long-Term Reductions in Energy-Based Carbon Emissions*. Ottawa: NRTEE.

Natural Resources Canada. 2008. *From Impacts to Adaptation: Canada in a Changing Climate 2007*. Available from http://adaptation.nrcan.gc.ca.

Paehlke, R.C. 2008. *Some Like It Cold: The Politics of Climate Change in Canada*. Toronto: Between the Lines.

Phillips, D. 1990. *The Climates of Canada*. Ottawa: Minister of Supply and Services Canada.

Ponting, C. 1991. *A Green History of the World*. Harmondsworth: Penguin Books.

Stern, N. 2009. *The Global Deal: Climate Change and the Creation of a New Era of Progress and Prosperity*. New York: Public Affairs.

TransAlta Corporation. 2008. *2007 Report on Sustainability*. Available from http://www.transalta.com.

UK Energy Research Centre. 2006. *The Costs and Impacts of Intermittency: Key Issues, Main Findings*. A UKERC Technology and Policy Assessment Report. Available from http://www.ukerc.ac.uk.

Vanderburg, W.H. 2000. *The Labyrinth of Technology: A Preventative Technology and Economic Strategy as a Way Out*. Toronto: University of Toronto Press.

Weaver, A. 2008. *Keeping Our Cool: Canada in a Warming World*. Toronto: Viking Canada.

White, R.R. 2001. *Building the Ecological City*. Cambridge: Woodhead Publishing.

Index